Bibliographic information published by the German National Library:

The German National Library lists this publication in the National Bibliography; detailed bibliographic data are available on the Internet at http://dnb.dnb.de .

Imprint:

Copyright © 2019 GRIN Verlag
Print and binding: Books on Demand GmbH, Norderstedt Germany
ISBN: 9783668939028

This book at GRIN:

https://www.grin.com/document/465312

Andrei Gladikov

Integration of Blockchain Components in the Electricity Balance Area Management

GRIN Verlag

Faculty of Environment and Natural Resources

University of Freiburg

Tennenbacher Straße 4

79106 Freiburg

Integration of Blockchain Components in the Electricity Balance Area Management

Master thesis submitted in partial fulfillment of the requirements for the Degree of Master of Science in Renewable Energy Engineering and Management

by

Andrei Gladikov

January 8th, 2019

Abstract

The increasing uncertainty of the power supply in Germany due to the surge in fluctuating renewable energy, power plant failures and accidental changes in consumption lead to the energy system imbalance. This not only threatens the system security, but also turns into financially punitive consequences for energy market players. Current market-based solutions and the German established data exchange platform with market rules 'MaBiS' do not foster a bottom-up collaboration in balancing groups to timely offset schedule deviations. The capability to modify and/or shift consumption of industrial or commercial end-customers as well as to adjust energy output of generation units gives a unique opportunity for organizing a shared platform for flexibility services exchange to mitigate the risk of paying imbalance tariff. This thesis is therefore aimed to estimate the severity of recent balancing costs fallen onto balancing responsible parties (BRPs), the quality of forecast incentive, and prove a blockchain-based secondary flexibility market with quota mechanism of trading activation rights an efficient and economically viable tool for BRPs for offsetting schedule deviations at short notice. After developing a prosumer-enabled model which is system integrated complying with existing balancing management, various scenarios of deviations exchange are emulated. The focus of this work is set to justify the sufficiency of strategic value in blockchain technology as the market underlying infrastructure and as a complementary layer for the established data exchange platform. This thesis identifies blockchain benefits and regulatory challenges for synchronization of multiple flexibility providers, creating and managing deviations-conscious balancer communities, resulting in energy balancing costs reduction, enabling innovative business models and additional revenue streams for balancing groups.

Keywords: balancing group management, balancing responsible party, electricity balancing market, imbalance price, energy blockchain, service model, prosumer.

List of Tables

List of Abbreviations

AEP – Ausgleichsenergiepreis

BG – Balancing Group

BBG – Blockchain Balancing Group

BIKO – Balancing Group Coordinator (TSO)

BGM – Balancing Group Management

BNetzA – Bundesnetzagentur

BRP – Balancing Responsible Party

BSP – Balancing Service Provider

DA(M) – Day-Ahead (Market); **ID(M)** – Intra-Day (Market)

DSO – Distribution System Operator; **TSO** – Transmission System Operator;

DR – Demand Response

EEG – Renewable Energy Sources Act

EIC – Energy Identification Code

EDI – Electronic Data Interchange

EPEX – European Power Exchange; **EEX** – European Energy Exchange

EWF – Energy Web Foundation

ESS – ENTSO-E Scheduling System

iMS – intelligent Measuring System

ISP – Imbalance Settlement Period

GCT – Gate Closure Time

MA – Moving Average

MaBiS – Marktregeln für die Durchführung der Bilanzkreisabrechnung Strom

MPO – Metering Point Operator

NRV – Net Regulated Volume

OTC – Over-The-Counter

P2P – Peer-To-Peer = **C2C** – customer-to-customer; **B2B** – business-to-business

PoA – Proof of Authority; **PoW** – Proof of Work; **PoS** – Proof of Stake

PR – Primary Reserve; **SR** – Secondary Reserve; **TR** – Tertiary Reserve

reBAP (AEP) – regelzonenübergreifende einheitliche Bilanzausgleichenergiepreis

RES – Renewable Energy Sources

RLM – Registrierende Leistungsmessung

ROI – Return on Investment

SFM – Secondary Flexibility Market

SLP – Standardlastprofilen

Contents

1 Introduction

Electricity is very difficult to store, it must be produced at the same moment when it is consumed. In other words, in every instance of time just as much power is fed into the electric grid must be taken from it [27]. Therefore, the balancing groups (BGs), so-called virtual energy quantity accounts, were established as the commercial solution and the key incentive for synchronising generation and consumption in Europe. In the BG all producers and consumers must report and maintain balanced schedules by submitting them to their Balancing Responsible Parties (BRPs) who send them to the TSOs before gate closure daily. The BRP is bookkeeping each feed-in and draw-off from the grid noted by each participant. All additions and deductions with the balance of energy quantity accounts must not leave any sums behind. BGs are expected to provide their best estimate in energy schedules that are composed based on generation and demand forecasts for each quarter hour of the following day, notifying how many in kilowatt hours of energy the BG will send out or take in. Yet what happens if that BGs get their forecasts wrong, e.g. the actual consumption does not match generation? In this case, the TSO aligns the imbalance between supply and demand by procuring different types of control reserves. The problem is that there are very heavy financial penalties [1] for having uncareful planning for BRPs. They have to offset unforeseen physical deviations from the registered schedules using balancing energy at cost of control reserve activation. Plus, the deviations are expected to grow more due to the inevitable coupling of sectors that utilizes the RES electricity in sectors like e-mobility, heating and industry [2]. This combined with an increasingly high number of distributed volatile energy assets and storage facilities will escalate the scheduling process by BGs. It could be financially painful for particularly smaller community groups if the cost of collective predicting their draw-offs or intakes of energy not right falls back onto its members. Therefore, this thesis elaborates on how to socialize this imbalance risk to ensure this will not turn out as risky penalty to a BG.

The blockchain technology provides multiple opportunities for BGs and BRPs to mitigate the divergence from notified trade schedules, and even turn this deviation into a profit and hence to maintain balancing commitment in such circumstances beyond control. The proposed project is aimed to prove the concept of blockchain infrastructure incorporated with current IT interfaces of balancing groups, and to carry out cost analysis at the BRP. The thesis will shed the light on the cumbersomeness of balancing electricity processes in the German power market as well as discuss the strategic behavior of balancing responsible parties focusing on the effectiveness of their economical incentive to improve the quality of balance forecasting. Based on the gathered findings, the business value of BRPs is presented, and key areas for savings as well as losses potential are revealed. Following that, possible local flexibility trading scenarios are emulated, a quota-based marketplace [3] steered by the BRP is proposed. Finally, regulatory aspects of applying blockchain in Balancing Group

Management (BGM) and specifics about aggregator role and demand response approach in Germany is discussed.

1.1 Problem statement

The extreme imbalance prices of reBAP [4] and currently inefficient imbalance policies applied in the German electricity balancing mechanism [5] cause yearly significant financial damage to BGs, at about €212 million in 2015, €150 million in 2016, €186,8 million in 2017 and €93,2 million for the first half 2018 [1] (discussed in *Chapter 3*). The high imbalance costs are accrued to BGs only because they cannot manage to fulfill their load-demand estimations precisely on time or due to unforeseen events. In other words, the lack of communication and marketing tools between BG participants and the BRPs as well as insufficiency of information on the system imbalance [6] rules out the possibility for bottom-up adjusting oversupply or undersupply cases in nearly real-time. Hence, P2P flexibilities trading needs to be considered to support BRPs in balancing their portfolio internally and commitments in wholesale markets with significant reduction of financial deviation penalties [6], [8], [9]. Since an online meeting venue for agreeing on the re-scheduling and trading where the BRP can manage its pool community or where balancing group members can interact with each other is not existent yet, it is therefore worth to emulate logical scenarios for flexible services exchange for the benefit of all interested stakeholders. Also, it is important to check how blockchain technology-conform the currently utilized data exchange for the schedule communication with market rules 'MaBiS' is.

1.2 Research objective and research questions

The project ambition is to provide a framework for developing a blockchain-based local marketplace that facilitates an active involvement of prosumers and local consumers to make use of the flexibility they can offer. The following are objectives of the project:

1. Provide the framework of established balancing market processes and scheduling mechanism ESS with market rules MaBiS in Germany
2. Elaborate on how imbalance prices are formed and what serves as an incentive for BRP to deal with forecast quality
3. Demonstrate the balancing cost analysis in Germany, clarify on the decision making of BRP between two markets: Balancing Market and Day-Ahead (DA)
4. Pinpoint on the areas in the BGM where the blockchain technology may find its implementation

5. Bring awareness on current German regulations in demand response programs and intricacies between aggregator and BRP, develop a solution for aggregating prosumer's flexibilities

6. Emulate potential scenarios of BRP portfolio optimization in an active manner using quota market approach with price signals

The research questions were developed based on practical experience and current state of scientific knowledge. Given the existing experience of establishing a P2P energy trading platform with the focus on providing energy flexibilities locally as well as taking into account profound studying of blockchain architecture variations [10], [11], [12], it becomes difficult to embark on the path of potentially beneficial blockchain shared database system. This urges the need for a systematic approach for blockchain components incorporation into BGM as well as its economical feasibility analysis. Apparently, this research gap should be covered to provide key actors with key insights on the mechanics of available energy markets as a primary tool for offsetting the imbalances and more innovative bottom-up enabling distributed ledger capabilities. It is therefore interesting to find out how a local flexibility market can be organized and whether there is a utility and commercial importance from it at all. To achieve the research objective, the following questions need to be addressed:

RQ I: Who are the Balancing Responsible Party (BRP)? How German BRPs are classified and what tools they use in daily operations for resolving imbalances?

RQ II: Do BRPs need to be balanced out even more? How urgent and severe the balancing problem in Germany? How strong the incentive and balancing group commitment?

RQ III: How to minimize imbalance costs of BRP with blockchain technology? Could it sharpen the planning BRP activities and provide new avenues for internal portfolio interaction?

1.3   Unique value of the project and scope limitations

Today commercial power provision is extremely inefficient and does not give any opportunity for the BG participants to rescue themselves locally from a shortage or surplus by offering their own flexibilities in a widely accessible and transparent manner [13]. Once a feature of blockchain-based P2P trading platform [14], [15] steered by BRPs is integrated into a BGM mechanism, new opportunities will become available for the group members to market flexible services in real time at the relevant time slots when supply imbalance arises in the power system. Whereas BRPs' advantage would be self-balancing portfolio optimization that results in a lessened need of going on wholesale intraday market for urgent buying the imbalance energy to fulfill TSO's request. Because of the large workload needed for this alongside with high price volatility that is de-touched from day-ahead prices due to a different pricing mechanism, the intraday continuous trading for many BRPs without internal flexible capacities is undesirable. This would also allow to avoid external pricey, time-costly and complex clearing and settlement procedures and other processes in electronic commerce involving many intermediaries [13] such as clearinghouse (ECC), exchanges (EPEX), brokers and index agencies (ACER). A shared marketplace attracts BRPs or power utility companies to managing price signals for its BGs.

The study could serve as an economical and engineering-scientific research, which considers most recent changes in economic incentive for BRPs in improving the quality of balancing as well as provides insights on the possible scenarios for organizing a secondary flexibility market utilizing underlying blockchain infrastructure layer. Based on such analysis the detailed models for integration of blockchain into existing IT systems of BGs and BRPs could be simulated. This research aimed at identifying what legal requirements need to be satisfied for successful adoption of blockchain-based customer-to-customer (or P2P) interaction under surveillance of a service provider in face of power utility companies, aggregator or 'BGM-as-a-Service' entities. Once the proposed approach is realized the advanced in economic efficiency processes of BRP could accelerate the promotion and deployment of RES projects as well as open up new avenues for marketing the network usage rights by large consumers in a decentralized fashion. The latter can significantly mitigate the threat of security of energy supply due to a sharper response to accidental consumption changes.

It is worth noting that the *Network Code on Electricity Balancing* (*NCEB*) – a European Target Model blueprint to harmonise all Member States' energy only markets' trading arrangements – got finalized, published and legally entered into force on 23 November 2017 [16]. However, due to the novelty of the enforced EU regulation and lack of any information on the German experience for adapting new guidelines, the Code will be set aside from the scope of the thesis.

Demand-Side-Management techniques to relax congestion management and detailed DSO role in the framework of German *traffic light* concept provided by USEF [17] is out of the scope of the thesis but also is worth noting because it presents a much larger savings potential. This could be explained simply by a 14-fold cost uptake in redispatch and curtailment of renewable feed-in since 2010 until 2017 [6], while the system balancing cost, on the contrary, have steadily been declined showing a 58%-drop over the past three years which is showcased and analyzed by the actual master thesis. Taking the urgency and severity of the problem into consideration, an innovative flexibility market concept where a transparent coordination ENKO-platform was recently proposed [18]. Accepting major added values of blockchain such as higher degree of decentralization, transparency and independence of third parties preserving users' data privacy and sovereignty aspects, this platform serves as a venue for flexibility requesters (DSO) and providers (consumers and producers as well as aggregator and BRPs). Based on the sensitivity prognosis prioritizing the flexibility providers for relieving network congestions, the concept was developed in the framework of research project NEW 4.0 for more efficient and active congestion management in the north of Germany.

Additionally, the third party aggregator model to enable end-consumers in face of smaller BGs to market their consumption and generation flexibility capabilities in regulation market is out of the scope as well, although it got recently legally approved [19] by the German regulator Bundesnetzagentur (BNetzA). Therefore, the part of control reserves is not covered in very deep details to not skew the reader focus, though only main split and features are provided.

1.4   Outline

In summary, the integration of components into balancing group management, namely the addition of a secondary market to the BRP toolbox as well as synergizing with MaBiS processes is thought to strengthen market commitments of BRPs and hence drive the electricity balancing costs down. Therefore, the actual thesis tries to justify the potential of blockchain premises in the field of BGM from different angles, thus structured respectively. Firstly, by thoroughly studying the target group – Balancing Responsible Party (BRP) – their daily operations, incentives and strategies *(Chapter 2)*. Secondly, by extensively estimating the severity of the balancing problem in Germany from the economical perspective *(Chapter 3)*. Thirdly, by overviewing the regulatory environment for end-customer P2P trading and by developing a model for its legal activation to expand balancing benefits *(Chapter 4)*. Fourthly, emulating scenarios from the internal portfolio optimization standpoint to enable a decentralized end-customer interaction *(Chapter 5)*. And finally, by finding blockchain incorporation fit areas in the MaBiS settlement processes from the technological point of view *(Chapter 5)*.

2 Fundamentals

This thesis seeks possible opportunities in the electricity BGM area for blockchain technology assimilation. In order to find these business cases, it is first and foremost important to extensively study the established processes for balancing in Germany, identify daily operations of market participants such as BRPs, and shed some light on the most recent regulations for electricity balancing. Followed by the economic evaluation of the most recent costs and incentive for balancing, the market application niche to enhance efficiency for balance group management is pinpointed. Afterwards, the most critical potential of blockchain technology is provided that suits to the described concepts. Deemed as a solution to strengthen the commitment of balancing group management, emulated scenarios for possible blockchain technology incorporation is outlined thereafter.

2.1 Key blockchain components to boost electricity balancing

Business models have significantly evolved due to economy's digital transformation. Today more than half the world's most valuable public companies have internet-driven business platforms. The problem with this model is that value generated by people has no equal redistribution among all those who have contributed to its creation, only large intermediaries who operate the platforms reap all the profits [20]. Furthermore, such traditional methods face a lot of bureaucratic hurdles related to time-consuming, corruptive and unreliable money transactions by using conventional banking systems [21]. However, a fast-emerging technology area called blockchain already affects the way energy companies do business and is to change this imbalance. Transactions, contracts, and their records are first to become completely public, unstoppable and immutable, without passing through any centralized middleman.

The claim that blockchain will revolutionize energy handling and redefine companies and economies has become ubiquitously spread [22]. It can already be met in other various fields, such as e-voting systems, carpooling, renting, land register, notary deeds, government departments, etc. [23]. Yet this thesis sheds light on actual potential of blockchain to change existing business approaches in electricity balancing area. The focus is on customer-to-customer dealings to strengthen balancing group optimization and showcase by emulating possible interactions over the secondary blockchain-driven marketplace. Finally, the thesis dispels blockchain's premises as a top layer for the German established electricity balancing-relevant data exchange functioning according to the market rules 'MaBiS'.

<u>**Definition**</u>

Blockchain is a distributed, digital transaction technology that enables secure storage of data and execution of smart contracts in peer-to-peer networks [24]. In other words, it is an open, distributed ledger that can record transactions between two and more parties in an efficient, verifiable and immutable manner. It can also be programmed to trigger payments automatically. Every party can verify the records of transaction partners directly, without an intermediary. With the blockchain technology, the embodiment of contracts in digital code is realized, contracts are stored in a transparent manner on the shared databases that are resistant to any data removing, tampering, and alternating. Potentially to every agreement, process and payment a digital record and signature are assigned that easily can be identified, validated, stored, and shared (Figure 2-1). As a result, intermediaries like lawyers, brokers, and bankers can be made redundant [25]. And individuals, organizations, machines, and algorithms seamlessly transact and interact with each other not involving the middleman. This sounds like an ideal rejuvenating solution for already worn out red-taped utility bookkeeping [26], [27], [13].

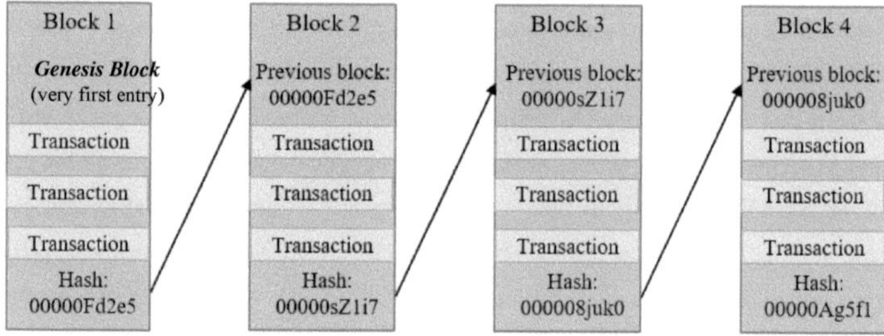

Figure 2-1: *Blockchain consensus mechanism* [28]

<u>**Historical overview on Blockchain**</u>

In 2008 the blockchain concept was initiated by an individual or group of individuals with unknown identity who call themselve(s) Satoshi Nakamoto [21]. They describe a protocol facilitating peer-to-peer transactions via a digital currency called Bitcoin *"without going through a financial institution"*. The blockchain movement appeared in the aftermath of the financial crisis due to general distrust of the commercial banking system [29]. It follows an ideology of destructuring established hierarchies, *"shifting societal influence from organizations to individuals, using democratic instead of autocratic decision processes, and empowering consumers claiming the high ethical standards of the new, decentralized world order"* [30].

Plus, according to the US-based consulting practice *Gartner* there is a game-changing potential for blockchain as it can enhance resilience, reliability and transparency in many centralized systems, as it stays in the middle of ascending trend of the *"Hype Cycle of Emerging Technologies, 2018"* [31]. This categorization shows the frenzy that surrounds Blockchain with its growing potential. Next, all important technical components of blockchain for this thesis are summarized in the Table 2-1.

Table 2-1: *Key technical characteristics of the blockchain technology* [32], [28]

	Definition
Central vs. decentral architecture	The task to be responsible for archiving and settling is shifted from central entity toward a number of individual entities. Because they possess their own copies of the ledger parties work in parallel not relying on the central server
A shared ledger database	A logical element allows seeing a respective ledger and interact according to ecosystem's rules; securely shared record of all transactions
Node application	Allows the participation in a blockchain ecosystem, has to be installed on the computational unit
Virtual Machine (VM)	Executes algorithmic codes, creates runtime environment for smart contracts, isolated from blockchain. Nodes running VM must execute the same instructions to maintain consensus across the network
Smart Contract	Electronic contracts – application on the Ethereum platform [33]. They observe persistent scripts, and are automatically triggered when certain events happen (once *if-then* condition is fulfilled)
Hashing	A cryptographic method of mapping input data by calculated unique output, guarantees immutability, a cornerstone of the blockchain technology
Consensus mechanism	**Proof of Work** – gives a user to the right to append the next block by solving mathematically-complex puzzles using CPU (mining), energy-intense process **Proof of Stake** – computational intensity by *PoW* solved by placing a deposit of digital currency in the blocks of system validators to increase system's chances to succeed **Proof of Authority** – modified *PoS*, involves the necessity for proving identity of the participant when joining the network playing a role of staking in lieu of the staking with the monetary value
Level of access	**Public** – any party or device can join to the network and submit transactions. Maintained by anonymous validators like miners or stakers who are irresponsible for the state or running virtual machine **Permissioned** – hosted by trusted authorities only who have rights to validate transactions, run virtual machine, and vote for the protocol changes. Customization options. Anyone can join this type of network once the certain role and identity is defined in an access-control layer **Private** - only *selected* nodes are given with the permit to read and access the blockchain ledger. This verified and authentic invitation is validated either by the network operator(s) or by the distinct set protocol ran by the network
Off-chain	Data is traced in private databases, public restrictions and higher transaction speed, scalable, feeless, retains privacy
On-chain	Irreversibly verifies the transfer of funds between two parties, reasonable for larger transactions, append-only, anonymous, higher degree of decentralization

| Tokens (cryptocurrency) | Internal means with specific attributes oriented around to transfer any value electronically over the blockchain network from one account to another (P2P) |

Unlocking the potential of Blockchain

Blockchain is a peer-to-peer network that sits on top of the internet. It was introduced as part of a proposal for *Bitcoin*, a virtual currency system that avoids a central authority for issuing currency, transferring ownership, and confirming transactions. Bitcoin is the first application of blockchain technology and can be defined as a digital bearer token. However, broad blockchain-based adoption still requires enormous institutional change. Blockchain adoption will take place by specific application forms and will depend on the degree of coordination needed to generate value and the novelty of the application. Widespread use in the future is predicted if companies start building up skills and learn how to change internal business model to adopt Blockchain capabilities [25].

One of the potentials of blockchain in the energy sector is to resolve process and transactions gridlock. Specifically, it could provide new foundations for many institutions (e.g. power utility companies and system operators) using applications reaching banking, insurance, legal or anything else that involves a binding contract [25]. Blockchain can be described similar to distributed computing (e.g. the TCP/IP stack – transmission control protocol/internet protocol), which over a period of forty years shaped the world wide web and the phenomenon of digital transformation. TCP/IP technology has also driven business model innovation – in just a few years coming from early adoption to digital ubiquity, dramatically changing the way businesses are handled.

By comparing blockchain to TCP/IP it is clear that as e-mail enabled bilateral messaging, blockchain enables bilateral financial transactions. Blockchain is open, distributed and shared platform for development and maintenance just like TCP/IP's. A disruptive transformation of energy sector might happen if blockchain dramatically reduces the cost of transactions and easily ensures the consensus among participants. Similar effect took place by TCP/IP when it unlocked new economic value by drastically lowering the cost of connections. Blockchain has the potential to become the system of record for all transactions [25]. If that happens, blockchain-based sources of influence and control emerge.

Today core function of the BRP business is to keep ongoing records of transactions, both financial and energy-based, to track past actions and performance and to guide planning processes. From this information they not only know how the organization works internally but also externally, organization's relationships. Normally, a BRP keeps its own private records, while TSO distributes received measurements across internal units and functions since they do not have a master ledger of all their activities [25]. To achieve agreement on transactions across individual and private ledgers takes a lot of time and is prone to error.

A typical stock transaction, for example, can be executed within microseconds, often not involving humans. However, the ownership transfer of the stock, known as settlement, can take several days. Since parties do not have access to each other's ledgers, it becomes impossible to automatically verify that the assets are actually possessed and can be transferred [25]. Instead, the record of the transaction crosses organizations and the ledgers are individually updated by a series of intermediaries who act as guarantors of assets.

By contrast, an interested party would host and maintain a ledger that is self-replicated by blockchain creating multiple identical databases. Whenever a change is added to one copy, the rest of the copies are updated at the same time. So, when transaction is forwarded, renewed information on the value and assets irreversibly enters in all ledgers. Hence, any necessity for third-party intermediaries to verify or transfer ownership is removed [25]. The beauty of blockchain-based system for an energy transaction lies in its settlement that takes place within seconds or minutes and in a secure and verifiable fashion.

However, scalability and security concerns, the cost of transactions, lack of supporting offline transactions and high hardware & resource requirements are major unresolved issues revolving around specific blockchain-based systems [34]. To cope with such constraints, various more flexible platforms have been developed (e.g. *Hyperledger* for B2B interaction [35], or *Tobalaba* by Energy Web Foundation [36], [37] for energy-related use cases). Due to the thesis scope limitations, the peculiarities of each particular blockchain architecture layer will not be studied but worth mentioning and left for future research and actual practical implementation.

In order to seize the global advantages of blockchain and accurately estimate the areas when it can be applied, the next section first covers all electricity balancing important aspects. It includes balancing market design, control reserves parameters, balancing responsibility as well as the incentive mechanism, roles and organization of the balancing group management are detailed. Alongside the content of the Chapters, the blockchain technology is always considered to be key to improve the related problem.

2.2   Balancing electricity and imbalance sources in Germany

A constant balance between the supply and consumption of electrical power must be guaranteed. Power imbalances cause frequency fluctuations in the system entailing equipment damage, infrastructure loss, and at worst, blackouts [5]. To maintain security of electricity supply, this responsibility belongs to TSO. Their job is to ensure that a control area they are responsible for is in equilibrium, both electrically and financially. In Germany there are four control areas and four TSOs respectively - TenneT, Amprion, TransnetBW and 50Hertz (see Figure 2-2). When imbalances arise in the power system, the participants of the balancing power market bid a price to alter production or consumption by open joint tender held by TSO. Rapid growth of variable renewable electricity generations from wind and solar energy sources created many challenges while integrating them into the power system. Because these sources are distributed and sensitive to weather fluctuations, forecast errors and power plants failure are common causes for deviations from load-demand schedules created by BGs, aggregated and managed by BRPs in energy systems. First, it is important to describe functions of BRPs and BGs.

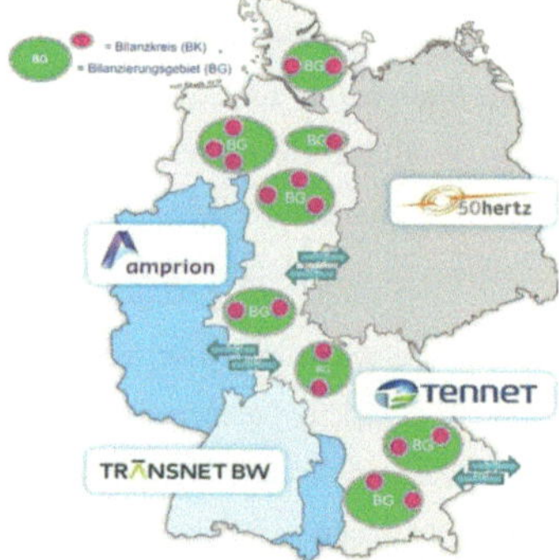

Figure 2-2: *Germany split into 4 control areas with schematic view of balancing groups* [38]

The imbalance settlement system is schematically represented below (see Figure 2-3). Foremost, it is important to keep in sync the entire system encompassing multiple countries in the EU. According to this, each country must maintain their grid state as balanced out as possible. Normally, separate synchronous systems are interconnected via HVDC cables (asynchronously), while an internal connection among control zones is built with AC cables (synchronously) [39]. For example, there is multiple count of synchronous areas in Europa, to name a few: the Continental European, the Baltic, the Nordic, the British, and the Irish synchronous area. Different control zones within one synchronous system aim to assist each other when disturbance in frequency occur. The individual countries ensure balance by encouraging BRPs with imbalance and real-time prices to link their feed-ins and draw-offs. Being a single buyer of balancing services, TSO usually offsets what is left after this matching process.

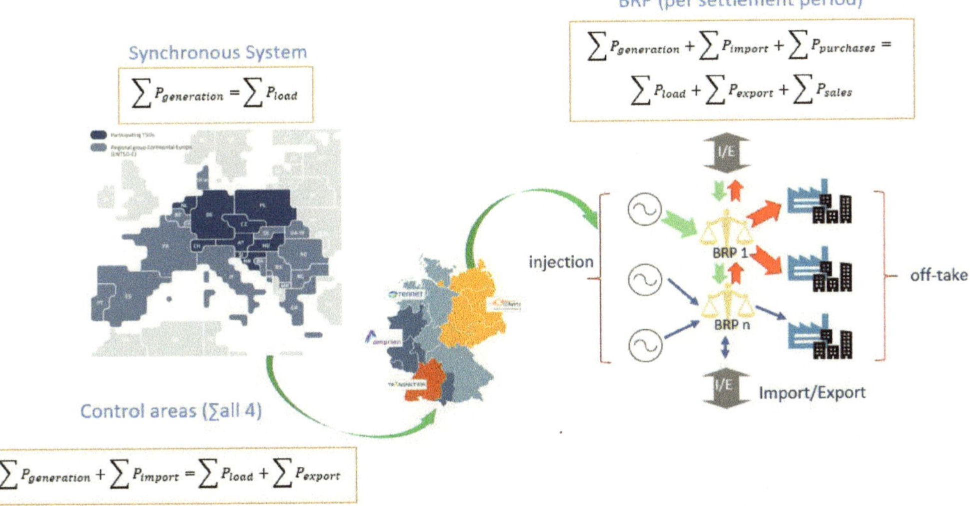

Figure 2-3: *Balancing responsibility in UCTE with TSO against BRP* [40]

A BRP portfolio that is made up of generations, energy purchases and imports on the left side of feed-in, and residential and industrial customers, energy sales and exports on the right side of draw-offs. By the TSO definition a portfolio will be balanced only when the following equation (1) holds true in each settlement period by their respective control zone [40]:

$$\left(\sum_{BRP} P_{generation\,(G)} + \sum_{BRP} P_{import\,(I)} + \sum_{BRP} P_{purchases\,(P)} \right) = \left(\sum_{BRP} P_{load\,(L)} + \sum_{BRP} P_{export\,(E)} + \sum_{BRP} P_{sales\,(S)} \right) \quad (1)$$

$P_x - capacity,\ MW$

When the wholesale trade ends by a point in time know as *Gate Closure Time* (*GCT*), each BRP declares its scheduled imports, exports and energy exchanges executed with other BRPs and established power exchanges. This type of schedules can also be called '*nominations*' and they have to be equalized in all four control zones in Germany and other adjacent EU member states. By doing so the balance between production and load guarantees the frequency stability which is a prerequisite from the perspective of power system security.

The mismatches at the BRP of submitted schedules with real-time measurements form short and long positions defined as the imbalances. In every control zone by calculating these imbalances the following equation (2) must be appreciated:

$$\left(\sum_{BRP} P_{G-measured} + \sum_{BRP} P_{E-nominated} + \sum_{BRP} P_{P-nominated} \right) - \left(\sum_{BRP} P_{L-measured} + \sum_{BRP} P_{I-nominated} + \sum_{BRP} P_{S-nominated} \right) \quad (2)$$

2.3 Balancing power market design

Whenever system imbalances occur within one imbalance settlement period (ISP), TSOs are taking over the responsibility not only for alleviating a single instance, but also for resolving the residual imbalances over several ISPs. To handle imbalance, TSO uses several balancing reserves, also known as regulated control reserves (Figure 2-4). Such reserves, which can act not only as suppliers but also as load and energy storage, can be differentiated according the technical characteristics, namely response time [41]. Any occurred frequency deviation from 50 Hz leads to the Primary Reserve (PR), automatic activation within 30 seconds across Europe. Once the system became stable, according to the common merit order list (*cMOL*) [42] the Secondary Reserve (SR) substitutes PR to roll frequency back to its standard value. the Tertiary Reserve (TR) is activated by TSO after prolonged usage of SR which capacity capability TSO wants to spare for next incidents. After the control reserves got utilized, the stage of settlement on costs from utilized balancing energy is fallen on the BRPs, which is the focus zone (highlighted in red) for this master thesis. Whereas the costs for capacity availability or holding are redistributed via grid fees to the end-consumers.

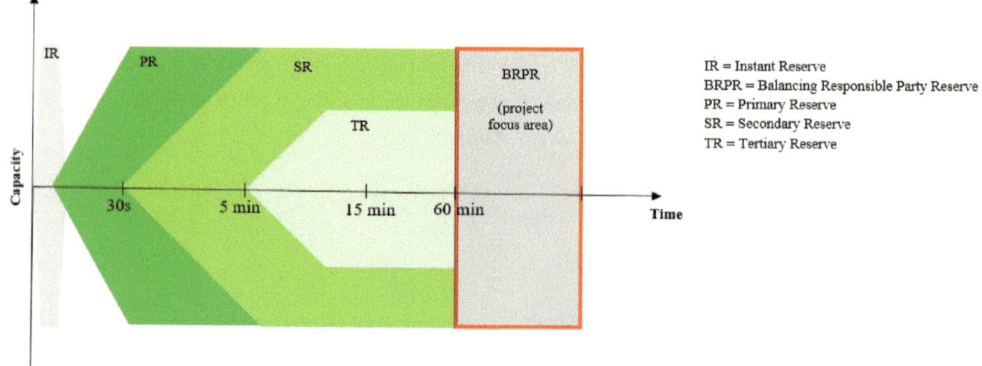

Figure 2-4: *Control reserve activation timeline* [43], [44]

The minimum capacity of all control reserves must be contracted by TSOs for the situations when local demand must be met after activation. The contracts of balancing capacity are tendered through auctions. It is important to note that in Germany a *pay-as-bid* system is used for control reserves such that successful *Balancing Service Providers* (*BSP*) are paid for capacity at the price (*Regelenergiepreis*) they bid into the tender [45]. Positive balancing capacity is used in case of the system shortage, while negative for resolving energy surpluses in the system. Technical requirements for bidders of balancing power are summarized in the Table 2-2.

Table 2-2: *Design specifics for the German balancing market* [5]

	Primary Reserve	Secondary Reserve	Tertiary Reserve
Auction type	Pay-as-bid		
Auction period	Week	Week	Day
Number of products	1 (base, symmetric)	4 (positive/negative; peak/off-peak)	12 (positive/negative; 4-hour blocks)
Minimum bid	1 MW	5 MW	5 MW
Reserved capacity	600 MW	2000 MW	2500 MW
Capacity payment	Yes	Yes	Yes
Energy payment	No	Yes	yes
Response time (ramp rate)	30 secs	5 min	7,5-15 min
Platform	www.regelleistung.net		
Buyer	TSO		

The demand for calling balancing capacity has completely inelastic character because it is not sensitive to the prices and is only defined by the imbalance state of the system. In other words, no price signals are given for electricity consumption, the electricity is consumed only then when it is needed [46]. For the submitted bids for balancing capacity there is a bidding ladder aligned in merit order where the least expensive bids have priority. The bidding ladder gets updated every four second in the course of imbalances occurrence followed by active matching bids with respective BSPs. Other situations when the BSPs use the imbalance settlement system contributing their imbalance positions in the opposite direction to the system imbalance are called *'passive balancing'*.

Once the control energy was utilized by TSO to restore the system balance in the respective control zone, the market-based accounting and settlement for involved BRPs happens after real-time (see Figure 2-5). If BRPs had misbehaved according to their schedules submitted before real-time, the energy deviations will be accounted with imbalance settlement price (*Ausgleichsenergiepreis*) in regard of BRP's individual balancing state and the balance situation of the system in a single settlement period. These deviations can be offset on the energy intraday market or via OTC, or by means of other decentral solutions like a blockchain-based marketplace, the focus of which the actual thesis was set upon and will be elaborated in next *Chapters*. Imbalance mechanism applied in Germany is non-discriminatory single pricing. This means that during settlement long (oversupplied BRP) and short (undersupplied BRP) positions are linked with one price [5]. The computing of imbalance prices is relied on the average costs of only the energy component of reserves activation.

For this stand only Secondary and Minute Reserves. The single pricing approach is a zero-sum game for the TSO.

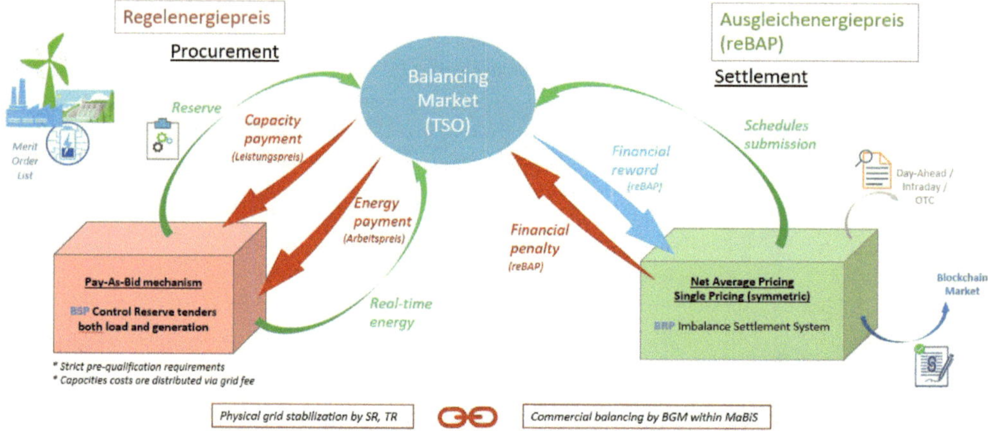

Figure 2-5: *Graphical representation of procurement and settlement with the TSO as a central role in Germany (own design)*

Further, it is important to define and describe functions of balancing groups (BGs) and their managers – Balancing Responsible Parties (BRPs).

2.4 Balancing Group Management (BGM)

2.4.1 Balancing groups

BGs are virtual energy quantity accounts that balance all the actual supplies and withdrawals (physical grid connections) and energy flows between other BGs (commercial transactions) within a control area [47]. For the efficient management of BGs, the *Energy Identification Code* (Y-EIC for DE national and X-EIC – EU-wide [48]) is generated and assigned to each of BGs by a corresponding TSO in coordination with their BRP. Alongside, the *Balancing Group Contract* is concluded between the BRP and the TSO [49], [50]. This contract contains several regulations on data provision for billing of BGs.

The coordination of feed-in of power plants is realized by means of load forecasts. Therefore, every producer and every load are contained in a BG. There are two types of energy consumers in Germany [51]. First is RLM (*Registrierende Leistungsmessung*) which represents detailed power metering with the energy consumption of more than 100 MWh el. yearly drawn, e.g. large industrial consumers. Such customers must be equipped with a specific electricity metering device with resolution of 15 minutes which can transmit 35040 measured times series a year to the DSO.   Second is SLP

(*Standardlastprofile*), take form of standard load profiles of typical residential and commercial consumers based on the past measurements. In Germany they are provided by BDEW (*Bundesverband der Energie- und Wasserwirtschaft*) and approved by DSOs at each control zone [52], [53]. For example, H0 for households, L0 – agricultural farms and G1 for businesses. To ensure that the consumption and production are harmonized in a control BG, the virtual concept of BRPs was introduced.

2.4.2 Balancing Responsible Parties (BRP)

The BRPs are main participants on electricity wholesale markets, they can purchase and sell electricity, either Over-The-Counter (OTC) or on energy exchanges. Traditionally, generation and system operation are decoupled meaning that TSOs do not receive any real-time information on production and load [54]. Therefore, BRPs provide the corresponding TSO with schedules that list the net energy trade a BRP intends to execute. A BRP can either source forecasted capacities from his own balancing group or externally purchase various electricity products on the wholesale markets. For example, long-term contracts on OTC for securing the base load, supply programs for the seasonal requests or short-term supplies from exchanges to cover the load peaks. Depending on planning accuracy and possible unpredictable results from production or consumption side, a smaller or bigger difference for balancing group remains.

Specifically, a load BRP forecasts its consumption in 1/4-hour increments for the next day at latest. Afterwards, this BRP purchases this energy from the other generation BRP or through power exchange. Both BRPs submit their trade to the TSO until gate closure by 14:30 in Germany [50]. TSO cross checks if schedules are matching and provides the BRPs with a positive or negative feedback in case of no error or sum/remainder was left over (3). Next, TSO compares booked schedules with meter readings provided by the DSO and bills any imbalances towards BRPs. The updates are allowed during the day 30 minutes ahead of actual delivery until the intraday cross-zonal gate closure.

$$Feed\text{-}ins\ Sums + Purchase\ Sum = Withdrawals\ Sum + Sale\ Sum \qquad (3)$$

A more detailed time flow of the BRP daily operations is schematically represented on the Figure 2-6.

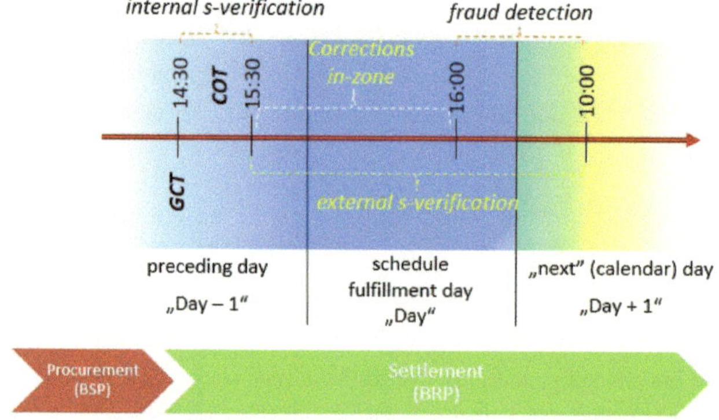

Figure 2-6: *The deadlines for the submission of schedules*

Before *Gate Closure Time* (*GCT*) the TSO procures control energy based on the bids for balancing capacity submitted by BSPs, while after GCT starts the process of settlement with BRPs. In terms of the settlement of the schedule registration for the fulfillment day D in Germany there is following breakdown of the fundamental phases (Table 2-3):

Table 2-3: *A breakdown in the phases for the schedule registration in Germany*

Phase	Description
A	Day-Ahead Phase: takes place the month before the D-1, 14:30 o'clock
B	Day-Ahead Matching: lasts 1 hour from D-1, 14:30 until D-1, 15:30 o'clock. During this phase TSO verifies internal schedules
C	Intraday Auction: reconciliation process starts at 18:00 o'clock, D-1 until the GCT of the respective delivery time
D	Subsequent schedules correction: begins at 15:30, *cut-off-time* (COT) of the respective intraday trading phase and ends at 10:00 am of the next calendar day. From 15:30 of D-1 up to 16:00 of D. Starting from only in-zonal corrections. Proceeding with fraud detection procedures [55] TSO performs cross-border schedules verification until 10:00 am of D+1.

Since the imbalance settlement period (ISP) in Germany has a granularity of 15 minutes, for each production schedule, there is a consumption schedule and vice versa. Based on this a BRP is obliged to pay for the established imbalance energy. According to [41] and [50] German BRPs contractually must stay balanced for each ISP and any intentional deviation is considered as infringement of duties. In fact, the difference between the imbalance price and day-ahead price is the penalty for having imbalance. Its value accounts for the aggregated deviations between scheduled and physical net energy during an ISP. Given that, almost all energy market players can be a BRP whether a utility, trader, consumer, and generator. The Figure 2-7 below differentiates BRPs into possible categories according to [56] and [57]:

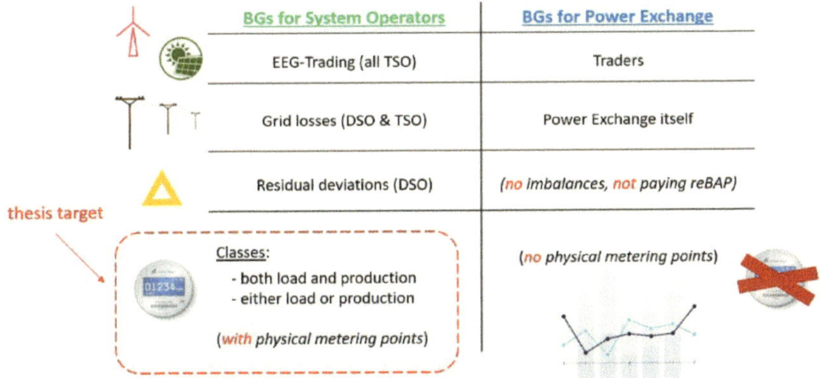

Figure 2-7: *BRP classification by specific purposes*

From the perspective of the German balancing group accounting model, further visualization of BGs and their data exchange is sketched for the sake of clearer understanding [57] (see Figure 2-8). Generally, the management of different balancing groups is separated into three categories:

- more than 100 000 customers
- fewer than 100 000 customers
- TSO (BIKO) plausibility check

An aggregate balancing group managed by DSO consists of several other balancing groups. Specifically, feed-ins and draw-offs from conventional consumers and power plants, then only feed-ins from RES (EEG), then from network losses and from a difference residual balancing group. The EEG, difference and loss groups are managed by the system-designated BRPs or by system operators themselves, whilst the feed-ins and draw-offs from conventional customers by commercial BRPs.

In case of smaller customer bundles the DSO allocates all values to a single *'Grid-BG'*, while for larger bundles it is split into individual BGs that system operator leads correspondingly. Finally, the BIKO has to ensure that all feed-ins and offtakes in the entire control zone will be allocated to a single BG. Therefore, BIKO does plausibility check of each balancing areas as part of the settlement with BGs. As a result, the TSO reports to BNetzA the final deviation allocated in the Delta-BG.

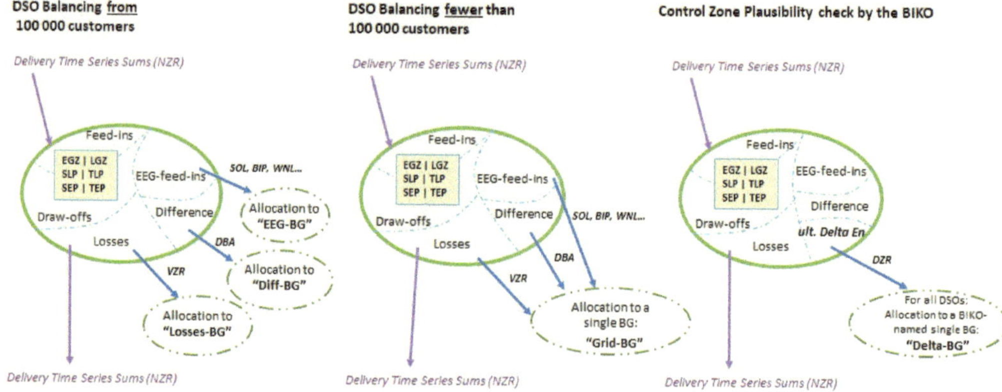

Figure 2-8: *Balancing group accounting model* [57]

Also, there is a wide range of a variety of metered time series coming in and out the balancing area controlled by respective DSO, either upstream or downstream. This is out of scope of the thesis and further detailed decryption can be found in the BDEW code list [58]. However, since it is important to know how the deviations in the energy system are revealed, the key organization of profiles flow is schematically represented in brief. First, a short note on the most important types of time series and insight on how they are managed is grouped as following (see Figure 2-9):

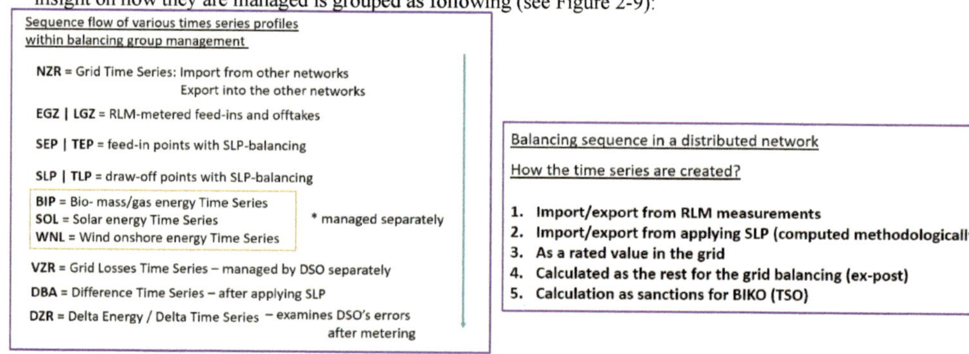

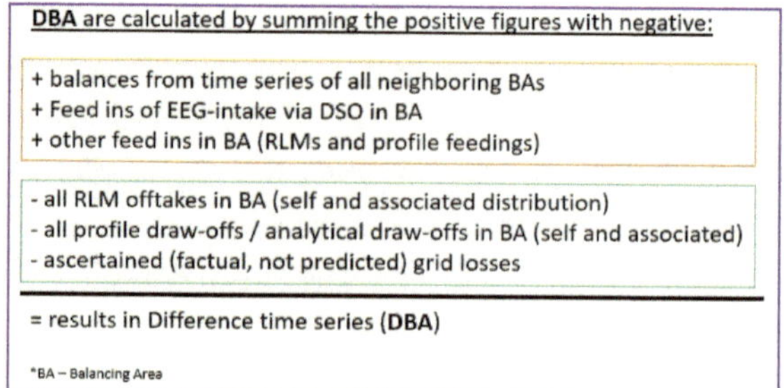

Figure 2-9: *Modular representation of processes in a variety of time series flow*

In order to determine the final balancing error (reflected in the profile *DZR*), DSO reconciles all aggregated metered values with seized profiles from several balancing groups. The following waterfall diagram (see Figure 2-10) demonstrates the whole balancing process of a distributed network dividing into three main stages [59]:

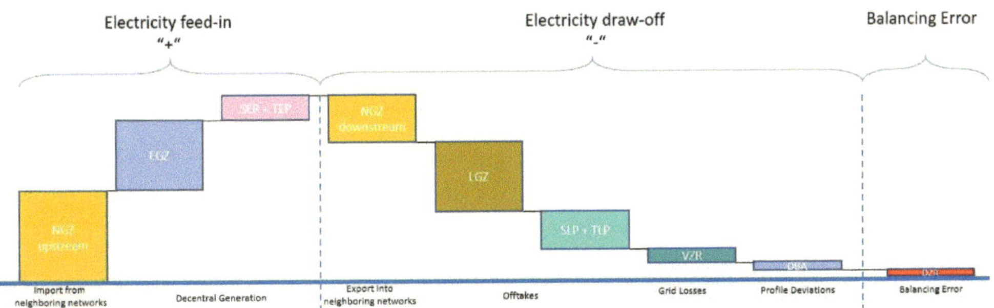

Figure 2-10: *A breakdown of balancing in a distribution network* [60]

As it can be seen from the bar diagram above the largest in size is the input of profiles from neighboring networks. Second largest are the feed-ins and offtakes from large RLM-metered customers. The most interest for the thesis present large industrial customers (RLM) since they have most significant influence on the final deviation from the schedules thereby threatening the security of energy supply.

To avoid confusion on several types of BGs based on the above demonstrated possible BGs accounting composition and energy profiles flow, the focus of the current thesis falls on the non-system, non-exchange and non-EEG BRP classified in 1) and 2) below in the Table 2-4:

Table 2-4: *Classification of BRP*

#	BRPs of different type
1)	BRPs that consist of *both* load and production. They have feed-in and draw-off metering points in disposal. Classically, this model fits to a standard utility that manages a power plant portfolio and provides provision services to its load customers
2)	BRPs that consist *either* of load or production. They deliver the entire production or demand to the market having a 'thin' relationship (between supply and demand) only in the form of schedule notifications. These are large industrial consumers or conventional power plants
3)	BRPs that do not own any metering points. Such groups are known as virtual profiles that limited to billing and accounting activities. This construct can be represented by mere traders and trading companies. Normally, they aim at zero sum of all their trades for their schedules

More generally, BRPs maintain balancing in the control area, namely they shall be responsible not only for a balanced quarter hour performance of the feed-ins and draw-offs allocated to their BG, but also stand for an appropriate schedule management and the economic balancing of remaining balance deviations. BRPs comprising loads could also be municipal, industrial and residential types. Usually, BRPs are created by various market players, such as utilities, traders, industrial consumers, renewable energy operators, aggregators, DSOs and even TSOs. For instance, the number of BGs accounts for 2435 with 543 BRPs only for TenneT, the largest operating TSO in in Germany [47]. Several BGs can be managed by one BRP. It is worth noting that BG and BRP were established only for commercial transfer of electricity, no physics and geography is involved. The spatial distribution of balancing groups does not play any role, whether they are geographically neighbors or many hundred kilometers away from each other, this will not bring any advantage to them. Only borders of control areas constrain BRPs geographically and not any network topology [61], [60]. Therefore, the efficient management and coordination of balancing groups is key.

2.4.3 Management of balancing groups

In order to reduce the exposure to regulation energy costs, BRPs must analytically control reactions of their customers in the BG. The following Table 2-5 contains important daily activities of BRPs [47]:

Table 2-5: *Key management activities of the BRP*

❖	Handing in load forecasts in the form of so-called *schedules* to the TSO. These schedules are composed from the consumer side (BGs), such as commitments from supplier, feed-ins from energy storage assets, also measuring data from energy trading firms. This procedure is iterated daily
❖	Requesting of actual consumption split into various consumer categories by DSO
❖	Billing of BRPs by TSO for balancing between actual consumption and production based on procured control reserve energy
❖	Contract conclusion and maintenance among BRP, TSO and DSO
❖	Handling the registration of BGs' Balancing Group Coordinator (in German - *BIKO*) und the corresponding DSO. The BIKO is basically a manager of individual balancing group
❖	Supervision of BGs according to the valid guidelines and regulations
❖	Billing of BGs regarding the *"Market rules for the conducting of Balance Group Billing Electricity"* (in German – *Marktregeln für die Durchführung der Bilanzkreisabrechnung Strom* – MaBiS) [38]
❖	Energy Data Management

The maximum efficiency of offsetting the imbalances is only achieved by aggregation of smaller customers in one portfolio, so-called *'pooling'*. By aggregating generators and consumers in the perimeter BRPs take advantage from the canceling out of individual imbalances. The emergence of flexible assets set up BRPs to merge into larger pools to increase the likelihood of matching opposite signs. Thus, the natural tendency for the maximum portfolio optimization is in expanding it infinitely by diversifying with as many as possible fluctuating energy assets [62]. Also, it enhances the chances of predicting consumer or producer's behavior and thus significantly diminish risk and uncertainty. The risk for paying penalties occurs when a BRP tries to optimize its positions to achieve higher profits and also being paid for this by the system.

Trading with counterparties can be categorized in a manner *'from the greatest to the least'*: exchange between neighboring EU-member states, adjacent control zones, within one control zone between balancing groups and finally within one balancing group (P2P). The following Figure 2-11 and Table 2-6 represent possible directions of communication for schedules on different levels between market participants in Germany [63]:

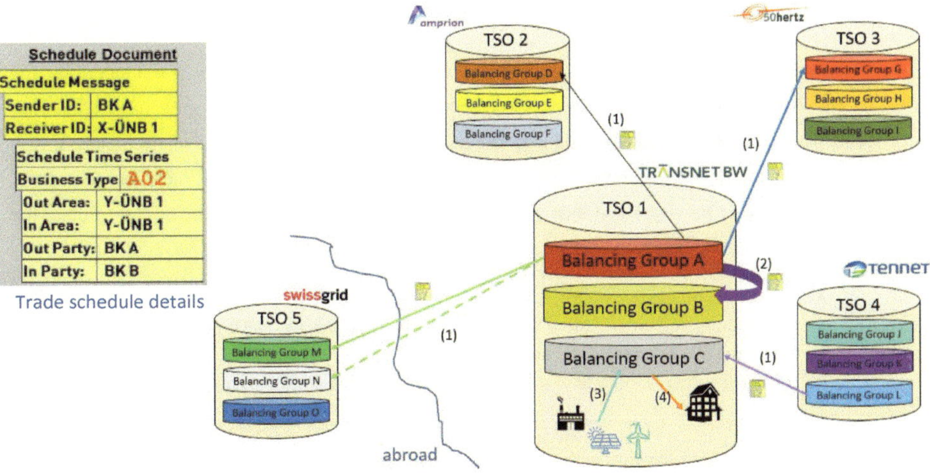

Figure 2-11: *Possible communication of schedules by business types in Germany* [63]

Table 2-6: *BRP classification by business types*

A		External
	(1)	Cross-zone within Germany and cross-border states businesses
B		Internal
	(2)	Businesses between BRPs within one control zone
	(3)	Production forecast (FC-PROD)
	(4)	Consumption forecast (FC-CONS)

The exchange schedule consists of feeding-in and drawing-off locations, electricity amounts and network usage time for each quarter hour of fed in or off taken including realized capacity exchange within the BGs. Of recent, in reliance on the *intelligent measuring system* (*iMS*) rollout, the 'smart' meter installation locations will be called market locations (*Marktlokation* - MaLo) and the balancing responsibility will be revised and set to re-assign from BRPs to the TSO by *01.12.2019* [64]. The measured energy values acquired by DSO with help of iMS will be transferred to TSO for further aggregation complying with the standards of GPKE and MPES. Subsequently, it is assumed a change in roles, specifically for metered data processing and transferring from DSO toward TSO in the framework of established market communication [59]. Further details are still under clarification.

2.4.4 Decentralized data management exchange framework in Germany

Currently, along with the German energy economic law (EnWG) exists the framework for decentralized information management designed by BNetzA describing the principles of market processes and rules for electricity and gas. The actual thesis considers only area of electricity. Under supervision by BNetzA following the decisions of German association of industry companies (BDEW), relevant messages and their format was defined and standardized. The uniform format for messaging is EDIFACT and currently being used for:

- Switching of suppliers (GPKE – electricity; GeLi Gas – gas)
- Accounting and settlement of balancing (MaBiS – electricity; GaBi Gas – gas)
- Metering (WiM – electricity and gas)
- Market processes for generating market locations (MPES – electricity)

The German EDIFACT framework is capable of handling around *40 processes*. In addition, the data management system in the electricity market is decentralized [65]. In other words, there is no central access point or platform for interaction, yet instead the *edi@enegy* model features decentralized data exchange based on bilateral communication between the market participants [66]. The Figure 2-12 below depicts the organization of the decentralized data management of German electricity market.

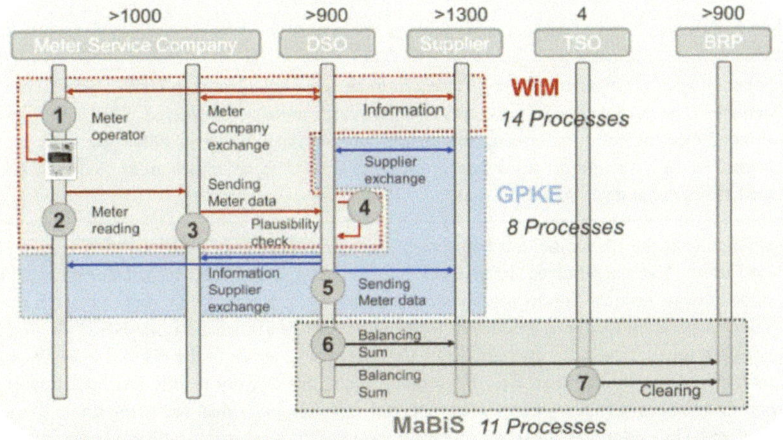

Figure 2-12: *Market roles and processes for data exchange in Germany* [65]

2.4.5 Balancing group accounting and settlement

As described in the *section 2.3.3* the TSOs stand for calculating the amounts of balancing energy that was utilized by respective BRPs within operating control zone, followed by the financial settlement. According to the German regulator BNetzA the required provision on the data exchange alongside with the obligation to cooperate in the course of the prescribed deadlines for this settlement process, known in Germany as *"Bilanzkreisabrechnung"*, is steered in line with the market rules "MaBiS" [60]. The most important steps are clarified below and visualized on the Figure 2-13 and detailed in Table 2-7.

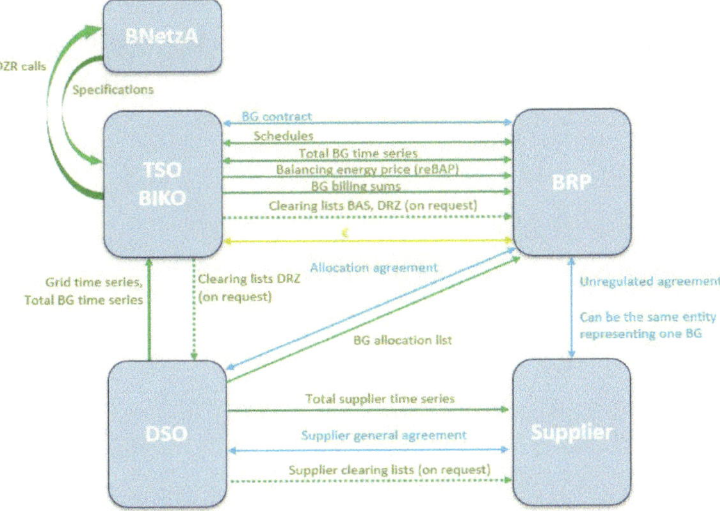

Figure 2-13: *Balancing Group Management model (adapted from* [67]*)*

Table 2-7: *Essential steps of the settlement process in Germany* [68]

I	Scheduling of all planned exchange transaction between BGs per each control area before the delivery. The deals beyond the borders of control area are also considered. Such schedules are shipped electronically undergoing the laborsaving automated process. After the delivery has begun, BRPs have right to adjust conformed with TSO their schedules up to 16:00 o'clock on the next calendar day.
II	DSO collects the 15-minute time series metering values of their grid-connected production and load units. The standardized profiles are also provided to the TSO by smaller consumption customers to account for in settlement procedure. Afterwards, DSO sums up all acquired values separating load and production types and ships the final data to the BIKO (TSO). The BIKO in turn redistributes the total balancing energy time series to the respective BRPs on the monthly basis. No later than the 29th workday ahead the delivery month, reconciliation of the energy balances between each network operator and each individual balancing group has to be finalized. Only after this, the data becomes relevant to the TSO and can be used to carry out the settlement of balancing energy. BIKO is obliged to issue the settlement not overrunning the 42th workday according to market rules.
III	Although the first settlement phase is over, there is a subsequent correction phase where the network operators and BRPs are allowed to submit to BIKO renewed schedules with necessary corrections. Taking this into account, the second stage after receiving corrected data lasts 8 months ahead the delivery month with fixed deadlines for clearing and data transfer outlined in the MaBiS.
IV	Up to the 20th workday after the delivery month the TSO is obliged to publish and electronically distribute the imbalance prices to BRPs to enable the settlement of balancing energy. This is performed already prior the reconciliation stage.

The invoices for allocated balancing energy are always issued by TSO toward the BRPs, regardless of the payment direction, whether a monetary sum is credited or debited by the party.

2.4.6 Scheduling and deviations

The electronic schedules matching and submission takes place at the central venue by TSO at each respective control zone. This process is performed in accordance to the regulations for *ENTSO-E Scheduling System* (ESS) [69], [70] with standardized data formats and names for market parties EU-wide. The scheme of the ESS contains four business types described above in the *section 2.4.3* and it is also divided into two parts: *'Day-Ahead'* and *'Intraday'* [71]. The Table 2-8 gives the breakdown for the ESS stages to clarify the order and approval in the BRP-to-BRP over the TSO communication:

Table 2-8: *The ESS scheme divide*

Day-Ahead ESS pattern	steps	Intraday ESS pattern
- Energy data attainment - Schedules creation by business type	(1)	Scheduling and Acquisition
Schedule transfer and feedback from TSO via *Acknowledgement* report (ACK)	(2)	Schedule transfer & ACK
Received schedules verification by internal business type: - check on the BRPs' submitted businesses - schedules' balances of both BRPs should match	(3)	Internal schedule verification
Schedules verification by external business type: - with other TSOs	(4)	Processing
Either *Confirmation* report (CNF) - all data is correct OR *Anomaly* report (ANO) - spots any inconsistencies BRP-to-BRP after fixing errors	(5)	External schedule verification
	(6)	CNF & ANO => *Final Confirmation* report

Furthermore, the following schematic illustration (see Figure 2-14) visually aids the importance of following the order in the schedule interchange between market parties:

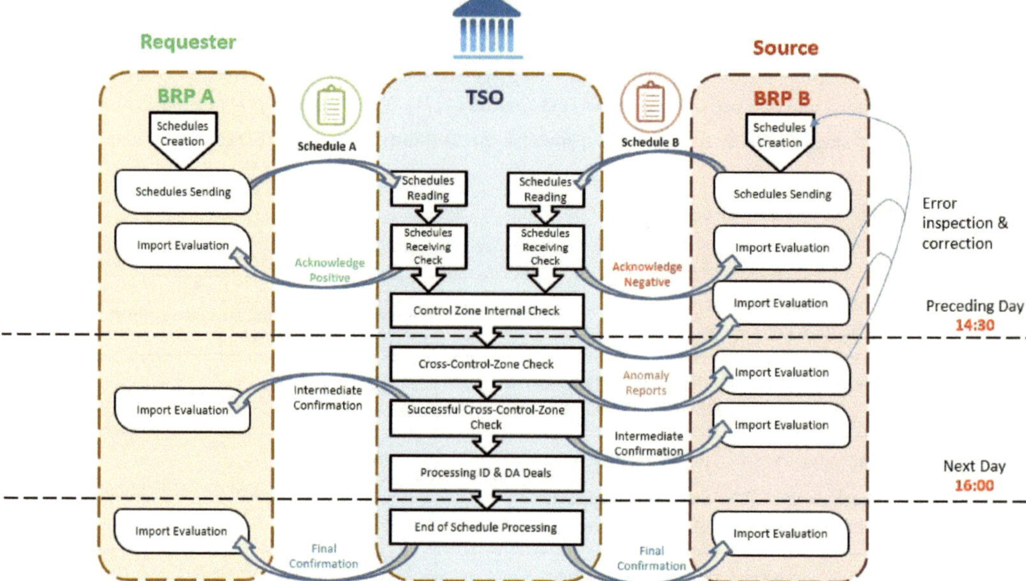

Figure 2-14: *Shipping of schedule with static and dynamic data, BRP A to BRP B over the TSO* [70]

Certain discrepancies within the schedules involve various types of errors that can be triggered by the ESS leading to a longer process of trade exchange between two BRPs. The Table 2-9 below lists the most important arising errors that could slow down the whole process and lead to the inefficient BGM:

Table 2-9: *The types of errors triggering one or the other type of report during the schedule transfer based on* [72]

Report type	Automatically triggered by
Acknowledgement report (ACK)	<ul><li>A formal error detection</li><li>Schedule was accepted even though it exceeded the deadline</li><li>Schedule was not accepted because it surpassed the deadline or Gate was not open</li><li>Quantity inconsistency</li><li>Control of line congestion and contract limitations – when these limits are not exceeded then the sum at the end should be zero and green light (or 'acceptance') is provided in the ACK report</li><li>Legitimacy and matching of the counterparty balancing group name and its EIC code</li></ul>
Anomaly report (ANO)	<ul><li>Value or time interval inaccuracies (quantities increased or decreased)</li><li>False counterparty registration</li><li>Counterparty time series missing</li><li>Time series not matching (modified)</li><li>Cross border capacity exceeded</li></ul>
Confirmation report (CNF)	<ul><li>Inaccuracy (mismatch with counterparty nominations)</li><li>Closure of the day-ahead processes</li><li>Closure of the day after processes</li><li>Registration of a schedule</li><li>New changes in the schedules</li><li>Gate closure was not reached yet (or passed)</li></ul>
Final Confirmation report (fCNF)	sent by TSO to BRPs by the next calendar workday by 10:00 am only after the correction phase closure. This report contains the data transferred from the scheduling system of the balancing group accounting

2.5 Imbalance settlement price in Germany

The deviations from submitted load schedules by BRPs are determined by the TSO based on the billing data available at the end of the 29th workday after the delivery month. Balance deviations are then allocated to the BGs or balancing sub-groups of the same contract. A balance deviation is basically an resulted difference between any or all withdrawals in one quarter hour allocated to the BG compared to any or all feed-ins in the same quarter hour allocated to the BG [50]. Once the balance deviation per quarter is sent from TSO, it is subsequently multiplied with a standard cross-control area balancing energy price (*reBAP*) [73]. The acronym stands for the German expression "*regelzonenübergreifender einheitlicher Bilanzausgleichsenergiepreis*". The symmetric imbalance price is provided in EURO per MWh and it has been a single uniform price for entire Germany.

In principle, the reBAP is composed of the costs of all 4 TSOs for the control energy activated (SR and TR) in a specific quarter of an hour and the associated control volumes in that same time interval. The reBAP is primarily calculated for each quarter hour block by dividing the total cost for the utilized balancing reserve energy by the total amount of utilized balance energy (4).

$$\text{Balancing Energy Price}$$
$$\text{Ausgleichsenergiepreis (AEP)} = \frac{\sum(\text{Costs} - \text{Revenues from called balancing energy}) \, (\text{€})}{\text{net imbalance volume (MW)}}, \text{€/MWh} \qquad (4)$$

For every quarter hour, there is a positive or negative reBAP due to the fact that control energy costs can be also positive or negative. A plus sign indicates when positive control (balancing) energy is procured (purchased) in the case of a shortage (undersupply) coverage across the control areas. And the low and negative values indicate the negative control (balancing) energy is utilized for resolving a surplus coverage across the control areas (oversupply). Next, it is important to understand different imbalance pricing rules.

For a fairer and more efficient imbalance price determination with many years of experience the BNetzA has applied several methods for the calculation of reBAP [74] (see Table 2-10):

Table 2-10: *Model for the calculating the reBAP by BNetzA*

AEP 1	The reBAP formula: net income (or cost) made with balancing energy settlement divided over the net activated balancing energy volume (known as *'Saldo NRV'* *). Negative sign of NRV translates into downward regulation, whilst a positive sign into upward regulation. At positive AEP1, BRP with negative portfolio balance pays to TSO, while the TSO rewards BRPs at positive imbalance. At negative AEP1, reverse is true
AEP 2	If the reBAP exceeds the highest (lowest) bid price for the activated control energy in a specified settlement period of 0.25 h, then the reBAP will be set equal to that bid price
AEP 3	In line of the sign of the balancing volume, upper and lower boundaries of reBAP are determined by the reference prices (average volume-weighted) on the intraday market at EPEX Spot. This method is known as 'capping' and 'flooring' and it eliminates any intentional or passive balancing by BRP portfolio optimization strategies
AEP 4	A surcharge or a cut is applied to the reBAP whenever 80% or more of the total negative (also positive) capacity was utilized by SR and TR to balance the system out. This is meant to incentivize BRPs more to manage their balancing groups more efficiently leading to lower future occurrence of such severe imbalances.

Saldo NRV – Net Regulated Volume balance

The reBAP price building mechanism being a weighted average of reserve energy utilization costs is not market-based since it does not reflect the marginal costs of balancing. This leads to inefficiently low average imbalance prices [5]. Similarly, the allocation of reserve capacity costs (PR, SR and TR) is also not fair since it is socialized to end-consumers via grid usage fees meaning that the market participants to blame in an actual imbalance are not penalized for their actions.

Overall, a central incentive of German electricity market design is to reinforce the obligation of BRPs to uphold set commitments, meaning that contracted generation and consumption will match actual generation and consumption. A more active management of balancing groups is achieved by high peak prices in the spot markets and imbalance prices (reBAP). It is extremely important for BRPs to know the system imbalance nearly real-time because it will enable them to react sharper on the occurring deviations suggested in policy changes by [5]. However, the BNetzA thinks that real-time publishing of reBAP can activate prohibited passive balancing that they believe is dangerous for the grid stability [75]. Nevertheless, this master thesis advocates innovative solutions that could facilitate the BRP portfolio optimization lowering the net deviation and lowering uncertainty and risk in forecasting.

3 Economical evaluation of balancing problem in Germany

This Chapter estimates how severe the balancing cost for BRP is and whether there is an urgency for strengthening BRPs' commitment to balancing. Essentially, trend analysis for latest imbalance prices in Germany is provided, followed with a short overview on the regulatory innovations in the field. The quality of balancing incentive for BRP graphically and numerically is determined and further used to find arbitrage opportunities between energy markets. Lastly, a country-wide financial cost development for balancing is demonstrated drawing conclusion with risk management strategies of BRPs.

3.1 Visualization of the latest consumption deviations for Germany

To understand how good in general the quality of forecasting by BRPs is and get the whole picture of electricity consumption in the entire country, the publicly available timestamped energy data in .csv format were studied. The used data source is smard.de/home/downloadcenter (*SMARD Strommarktdaten, Bundesnetzagentur*) [76]. The energy values are provided at the platform in quarter-hourly time resolution and in energy units MWh. For the observed period the entire 2017 year was chosen. On the platform there is two categories of available energy values: prognosed consumption and realized consumption. Both types were downloaded and analyzed for the whole year (Table 3-1).

Table 3-1: *The count of downloaded time series*

number of **forecast** electricity consumption values	35040
number of **realized** electricity consumption values	35040

The Figures 3-1 and 3-2 show how the forecast mismatches the real consumption. The period was split into two equal parts, each 6 months, for a more convenient look.

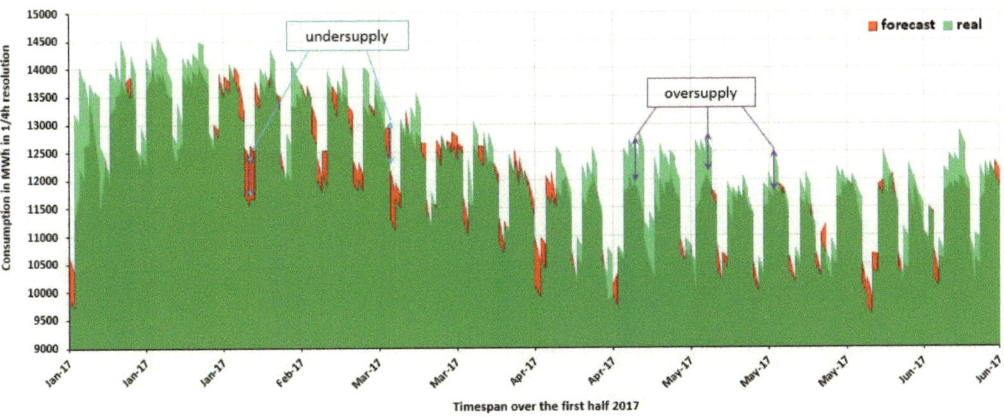

Figure 3-1: *Load forecast and realized profile visualized for the first half of 2017 in Germany*

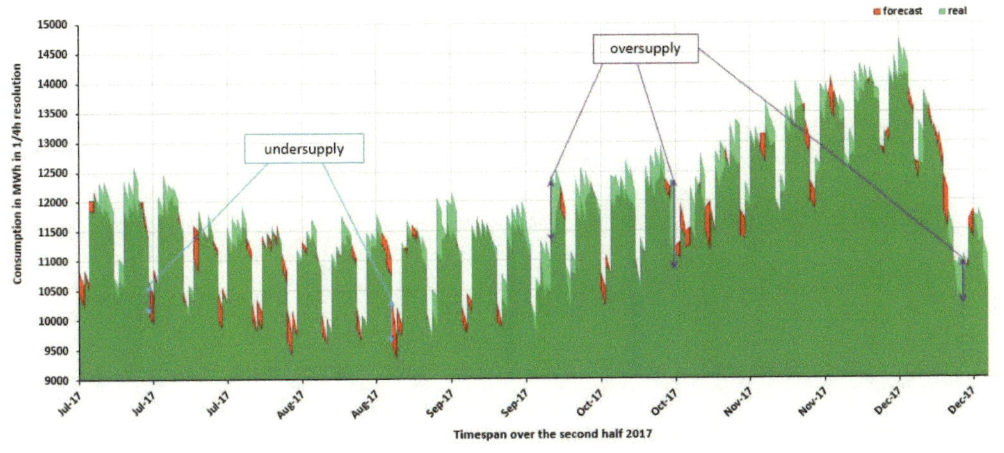

Figure 3-2: *Load forecast and realized profile visualized for the second half of 2017 in Germany*

The actual consumption values were overlaid over the forecast data. By doing so the mismatches and further on deviations that occur due to the difference between prognosis and actual measured on-site electricity can be revealed and estimated (see formula (5)).

Load-demand deviations = forecast value – measured values, MWh

$$(5)$$

Red painted areas stand for the situations of undersupply, meaning the real consumption turned out to be lower than initially scheduled one by BRPs. This indicates the state of shortage and therefore, holds name of *'short'*. Conversely, light green areas on the graphs illustrate the situations of oversupply which means there was an excess or real consumption happened to be higher than BRP estimation. It holds the name of *'long'*.

Qualitatively, one can notice a seasonal change in the pattern:

- Increased load demand in winter term from October until February due to the need for heating and intense operation of industry, and
- Decreasing consumption profile from May towards August in warm weather in summer

The measured load peaked at 19870 MWh per 15-minute time slot was observed once in 13th December and in range 19440-19520 MWh thrice in January. Noticeably, a significant gap in consumption could be seen in the end of December and the beginning of January which is explained by country-wide Christmas holidays during which most of large industries are out of operation. Whereas the minimum observed consumption values are in area of 8390-8420 MWh per 15-minute time slot in August and approximately 8600 MWh per 15-minute time slot in the month of June. As a result, the graph for real deviations was plotted (see Figure 3-3).

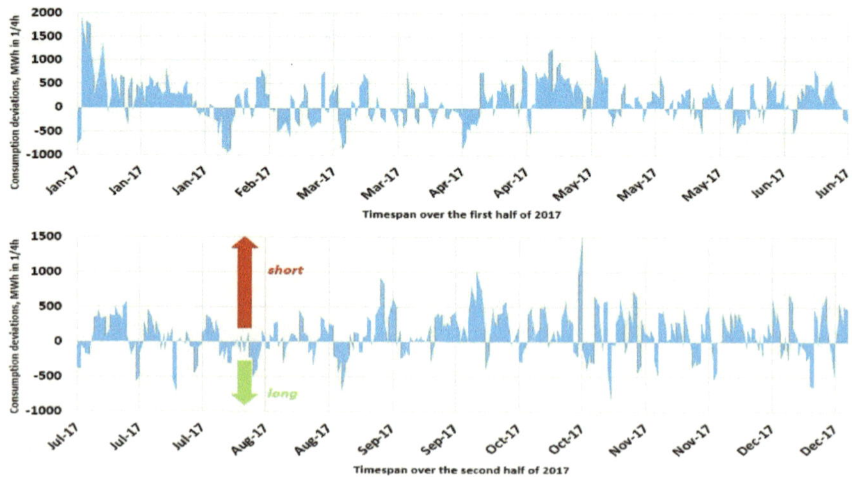

Figure 3-3: *Electricity consumption deviation in Germany, 2017*

The positive part of the surface diagram stands for the system shortage or undersupply, while the lower negative part translates into oversupplied occasions or excess. Markedly, the situation of when system was short took place more frequently than the long cases. The statistics of studied values were calculated using a simple conditional python script, and the result is as following (Table 3-2):

Table 3-2: *Statistics of the electricity consumption deviations in Germany, 2017*

How often?	Average	Total	Number of values
<u>System short</u> 63,25% of total observations	410,01 MWh	9 GWh	22163 positive deviation values
<u>System long</u> 36,75% of total observations	-299,13 MWh	-3,9 GWh	12877 negative deviation values

According to [62] the events when BRPs are short (undersupply) is considered mostly as profitable because less electricity was consumed and hence less fuel was burned alongside with activation costs of power plants. In contrast, during 'long' events BRPs bear losses. Therefore, from the charts above, it is apparent that with profit to loss ratio 9:3,9 the system BRPs had profits and tendency to undersupply, thereby spared on the electricity production.

The sudden change in the direction, from undersupply to oversupply, happens when there is a need to ramp up or ramp down the demand to stabilize the energy system. Roughly, when the frequency fluctuates below 50Hz the TSO requests to decrease the load, which means events of undersupply, while it surpasses the set limits above 50Hz the TSO requests down regulation or increase in load.

3.2 Latest trend of reBAP prices in line with regulation changes

In order to estimate how large imbalance cost for a BRP at a particular imbalance settlement period (ISP) might be, the historical development of reBAP values over the full 2017 year and the first half of 2018 was studied (Table 3-3). The prices were downloaded from the public German TSO platform regellesitung.net [77] in .csv format. On the Figure 3-4, the German imbalance prices are compared to the net regulated volume (NRV), which is known as *Regelsaldo* and means how much balancing energy was used by activation of secondary and minute control reserves to stabilize the system per each 0.25 h.

Table 3-3: *Count of the studied reBAP values*

Number of reBAP values for 2017	35040
Number of reBAP values for the first half 2018	17372

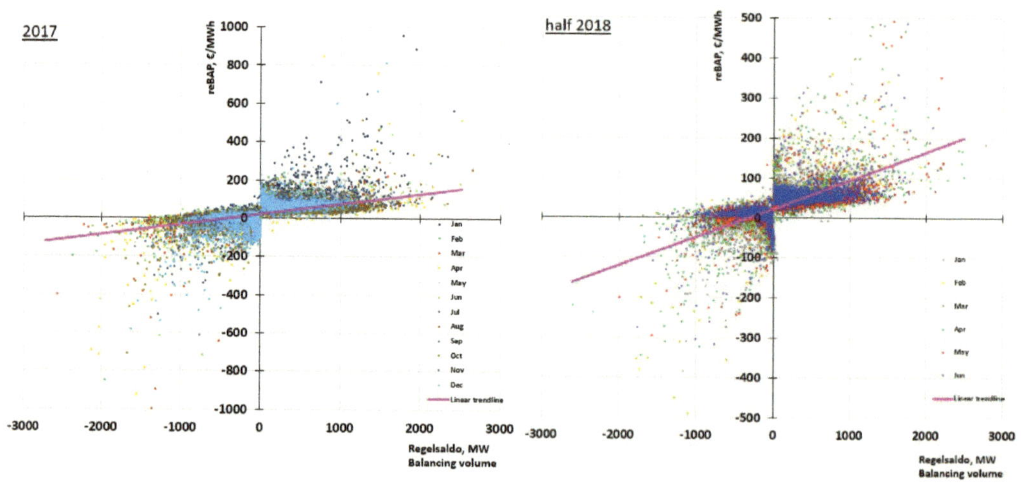

Figure 3-4: *The yearly historical imbalance price range in relation to balancing volume in Germany, 2017 and first half 2018*

By analyzing market values, the following information which is most recent and critical to the BRP billing processes was discovered (Table 3-4, Table 3-5):

Table 3-4: *Final BRP limiting economic metrics after control energy activation in 2017*

Extremes	"Most expensive" 0.25 hour for BRP	Activated control energy & imbalance price	Total monetary cost
Maximum Positive Settlement Price +24 455 €/MWh	+4 858 142 €	795 MW 24 455 €/MWh	€190 million
Minimum Negative Settlement Price -2 558 €/MWh	-10 509 €		

Table 3-5: *Final BRP limiting economic metrics after control energy activation in 2018 half*

Extremes	"Most expensive" 0.25 hour for BRP	Activated control energy & imbalance price	Total monetary cost
Maximum Positive Settlement Price +1 359 €/MWh	+691 345 €	2 035 MW 1 359 €/MWh	€94 million
Minimum Negative Settlement Price -1 873 €/MWh	-8 745 €	-800.5 MW -43.7 €/MWh	

Findings explanation

The findings of the current work and presented in [4] confirm that there could be extreme undersupply costs for BRPs who deviated at the historically critical imbalance events that took place on 17th October in 2017 during the time period 19:00-20:00. The estimated price for this hour could be *113 317 €* [4], which might be insanely high for smaller balancing groups with little capital, no price signals and proper risk management tools. Such abnormal imbalance price was explained in [78], and main reason why this happened was the absence of free bids in the German balancing market that in turn could limit the competition for secondary reserve [45]. This loophole has made contracted market participants strategically place zero-euro balancing capacity bids to win the merit-order list (*MOL*) with ridiculously high balancing energy price of 77 777 €/MWh.

Similarly, the energy values were analyzed for the first six months of the year 2018 in order to see the historic development of the reBAP (see Figure 3-4) and the effect of the recently applied measures by regulators for dealing with extreme imbalance prices in the German balancing market. Specifically, since 5th January of 2018 according to BNetzA the upper/lower limits for balancing energy (*Arbeitspreis*) were set to *(–)9 999 €/MWh* which implies no more extreme imbalance prices can be taken for imbalance costs calculation. For BRPs this means – rather far less size of extreme 15-minute imbalance sums computed from the price component, while the other component of deviated volume is still influential and must be prioritized first to minimize the end imbalance settlement cost.

Furthermore, recently on 12 July 2018 the German balancing market regulator BNetzA is developing further at rapid pace and tried to test again some new changes. Namely, the BNetzA introduced temporary a new mixed cost structure where the balancing energy and capacity prices (*Arbeitspreis und Leistungspreis*) merged into one bid on the balancing auction [79]. The aim of this new regulation was to increase the competitive pressure on the balancing energy prices in terms of the procurement of control energy, and thus make the procurement system more efficient. Indirectly, such measure threatens and disincentivizes future clean-tech branch that especially foresees an electricity market consisting predominantly of flexible decentral generators, new storage technologies and flexible consumers. At the same time, it can serve as a gift for the conventional power operators.

The enacted regulatory measure entailed an increase in capacity prices with strong volatility and a steady decrease in balancing energy prices. This master plan of the provisional mix-cost structure designed by BNetzA was meant to test the potential effect in the expanding of financing sourcing for conventional power plant bidders through the enhanced capacity prices and hence higher filed capacity costs. The capacity costs accrue to the activated control energy by secondary reserves are redistributed via the grid fees on the electricity billing of end-consumers, especially large industries and businesses can get an exponentially strong surcharge on their grid fee component in the bill [79].

However, after an urgent class action filed by *Next Kraftwerke* to BNetzA this experiment was stopped in two days by 14 July 2018 [79]. The pricing mechanism was set back to normal balancing energy price and capacity prices in separate blocks. The court hearing was suspended until 15 October 2018 due to the insufficient understanding on what measures the regulator must apply. Yet the following list of consequences (Table 3-6) if the mixed pricing approach would have lasted further were provided by *Next Kraftwerke* as serious dangers in the energy branch [79]:

Table 3-6: *Possible negative impact to energy sector after enforcement of mixed pricing approach for balancing market*

(1)	Additional costs for grid users
(2)	Reduced security of supply due to a weakening effect of reBAPs – the price is calculated with activation costs of balancing energy. As the balancing energy prices fell significantly in the course of the experiment so did the reBAPs. Low reBAP values translate into the weak incentive to BRPs for improving the quality of scheduling.
(3)	Complication and even can disable the integration of new technologies in the regulation market due to discrimination against providers with higher variable costs, e.g. RES

Anyways, had this regulated action of mixed pricing procedure taken an extended effect and had not been blocked, no direct negative impact on the imbalance billing for balancing groups could happen. In opposite, the reBAPs would have had lower values leading to the smaller imbalance settlement costs and hence weak incentive for BRPs. Fundamentally, proposed approach of mixed pricing has an unfair character because the BRP actually cause the fluctuations in the grid by having multiple forecast errors in planning and for this they are ought to be more responsible and not the end-consumers.

3.3 Juxtaposition of historical reBAP values vs. spot prices

One of the important decisions a BRP has to make is whether to take part in the correction phase on the traditional centralized energy markets, in Germany it is the Intraday Call Auction at EPEX Spot, or wait until the imbalance price gets published in 20 days after the delivery and pay *post factum* for their deviations. Also, the BRP can swing to another decentral market options like emerging blockchain-based decentralized markets by which means a BRP could fix *ex post* his forecast errors in an eased way. For example, an OTC-like P2P energy trading with blockchain infrastructure is already provided by the *EnerChain* project as a legal entity [80].

BRP's job is always to have a comparison among diverse sources of energy and flexibility service offers at disposal and follow risk-management strategies for avoiding unprofitable trades. Sometimes, the hesitation to adjust a position a BRP made before the delivery can actually be counterproductive and aggravate the system imbalance resulting in an incurred financial penalty. This is because the imbalance price (reBAP) is not known in advance and practically it is a zero-sum game for a BRP aiming at the most neutral portfolio state as possible. In trying to make additional profits by guessing and modeling the imbalance price development, a BRP can end up in a wrong direction bearing considerable losses.

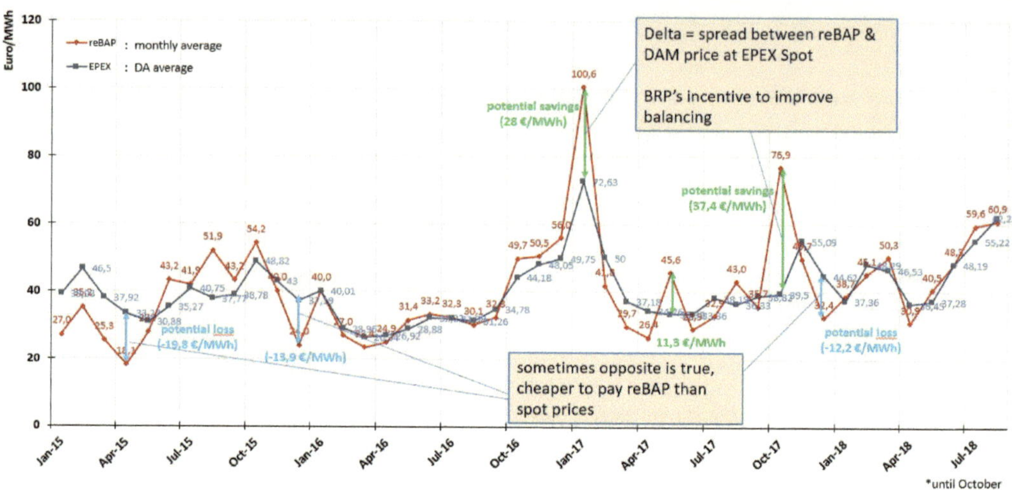

Figure 3-5: *Decision making rationale of BRP for portfolio adjustment* [81]

In fact, the main trading indicator for a BRP to improve their forecasting methods is the *guarantee of return on investment (ROI)*, which is determined by the imbalance spread (see Figure 3-5) [5], [6], [82]. The optimal behavior of a risk-averse BRP with no insight on the state of the system imbalance is to aim for minimization of its own portfolio imbalance only in case the costs for doing so are lower than the expected costs from TSO allocated through the imbalance settlement system *ex post* [62].

The imbalance spread (delta on the graph) is the difference between spot market price (Day-Ahead or Intraday Auction) and the imbalance settlement price (reBAP in Germany). For simplicity the prices are provided in monthly average resolution. From the graph it can be seen that the positive spread (when reBAP is higher than DAM price) usually results in potential savings for traders, while negative spreads when reBAP is below DAM price – in potential losses. Being on a right or wrong side at the system imbalance determines whether the BRP earns or pays the imbalance spread respectively. [5] When this spread is wide enough, it means the BRP is financially motivated and can see the potentially very high returns at the moments when they have positions counteracting the system imbalance. In essence, sizing the cost for balancing is mirrored by the imbalance settlement prices.

To better understand and visualize the current incentive of German BRPs to balance their portfolios, the imbalance spread moving average (MA) curves for years 2017-2018 were computed. The Day-Ahead spot prices downloaded from epexspot.com were taken as the comparison to reBAPs downloaded from regelleistung.net. The graphs on the Figure 3-6 were plotted using Microsoft Excel. In accordance to [5] and [83], the spread between reBAP and DAM prices is juxtaposed against the *net regulated volume (NRV)* to get insights on:

- how frequent the BRPs were short or long
- numerically estimate the imbalance spread size
- determine the quality of the incentive to instigate BRP own balancing

- shed the light on the distortion effect on the BRP incentive

The *NRV* stands for the difference between activated up-regulated volume and down-regulated volume per each ISP. The MA curves were calculated by sorting the imbalance spread values by increasing NRV with the interval parameter set to 180 ISPs.

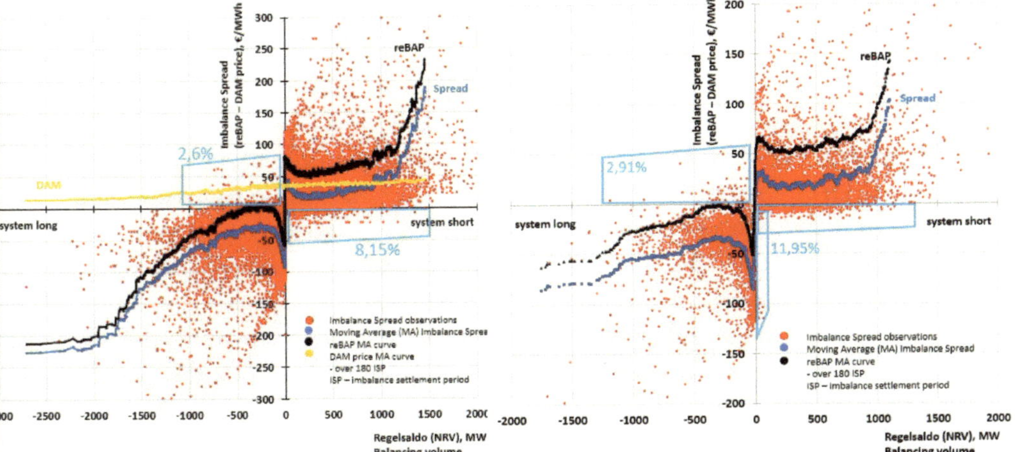

Figure 3-6: *Imbalance spread relative to system imbalance over 2017 and first quarter of 2018*

Positive requests ('short') for balancing capacity are made when there is a need in up-regulation by the system, while negative requests ('long') at the events of down-regulation. The next formulas (5) were applied by calculating the imbalance spread for two system states (Table 3-7):

Table 3-7: *Imbalance spread formula*

Negative calls	Positive calls
$\Delta_{long} = \lambda_{(-)reBAP} - \lambda_{Spot\,DA}$	$\Delta_{short} = \lambda_{(+)reBAP} - \lambda_{Spot\,DA}$

$$\Delta - imbalance\ spread;\ \lambda - price$$

(5)

Results interpretation

First of all, the analysis shows that with growing NRV the imbalance spread increases alongside. This strengthens the BRP incentive to reduce their portfolio imbalance. Secondly, as long as the reBAP MA curve (black color) lies above the DAM price MA curve (yellow color) in the events of upward regulation and vice versa by downward regulation, it means that the system remunerates flexibility [83]. On the both sides the reBAP curves have a right direction translating into the sufficient economic incentive for market players. The width between reBAP and DAM MA curves indicates how strong the BRPs are incentivized to balance their portfolio prior to the real-time. The wider the spread, the more likely this will drive up the demand for flexibilities on the various intraday markets and subsequently the price offers because more and more BRPs will want to offset their deviations not through the established imbalance settlement system.

However, if a decentralized market is used for ridding of deviations, it may impair the price discovery on the main centralized exchange because the BRPs will place their offers to a lesser extent if the right and better conditions are provided on the secondary flexibility markets. Also, at lower reBAP prices BRPs will be willing to passively wait the real-time imbalance prices leading the TSO to activate their control reserves.

In 2,6% of the left half chart the imbalance spread provided a wrong incentive for BRPs – it was positive at the state of system oversupply, while 8,15% negative – in events of system undersupply (right half) (Figure 3-6). Or from the combined perspective, in 5,9% of all imbalance spread values observed in 2017 convoluted the incentive for BRPs, which got worse by about 5% after similar performed analysis in [5]. This implies that it became harder and costlier for BRPs to not correlate with the system imbalance. Subsequently, similar computation was done for the first quarter of 2018, which shows a significant increase by 3% in perverse incentive in events of system undersupply, though the studied period is four times shorter than reference values of 2017. The Table 3-8 summarizes the results above.

Table 3-8: *Accuracy of the incentive for BRP provided by the system*

	Wrong incentive		
	Quadrant 1, occurrence	Quadrant 4, occurrence	Total for both
2017 (full)	2,6%	8,15%	5,9%
2018 (quarter 1)	2,91%	11,95%	9,15%

In analyzing the price values the following statistics (Table 3-9, -10, -11) were calculated by a python algorithm and Excel to determine the recent incentive for BRPs to start improving forecast quality:

Table 3-9: *Number of studied values according to the sign of NRV*

	The number of imbalance spread observations in 2017	The number of imbalance spread observations in first quarter of 2018
short upper right part	19 552 (55,8%)	5327 (61,68%)
long lower left part	15493 (44.,2%)	3309 (38,32%)

The total number of price values (Day-Ahead market prices, reBAPs, NRV) for 2017 (full): *105 120*

For the first three months of 2018: *8636*

Table 3-10: *Statistics on the BRP incentive for improving quality of forecasting, in 2017 (full)*

	Moving Average curve of reBAP prices	Moving Average curve of DAM prices	Moving Average curve of imbalance spread
Average positive price	62,2 €/MWh	35,8 €/MWh (NRV > 0)	26,4 €/MWh
Average negative price	-12,3 €/MWh	31,5 €/MWh (NRV < 0)	-43,8 €/MWh

Table 3-11: *Statistics on the BRP incentive for improving forecasting quality, in 2018, 1st quarter*

	Moving Average curve of reBAP prices	Moving Average curve of DAM prices	Moving Average curve of imbalance spread
Average positive price	59,15 €/MWh	36,2 €/MWh (NRV > 0)	22,97 €/MWh
Average negative price	-12,74 €/MWh	32,7 €/MWh (NRV < 0)	-45,43 €/MWh

Looking at the values of average prices it can be concluded that BRPs are twice as much financially incentivized in the 'long' area as in the 'short'. This could be explained by lower cost of a negative request in comparison to a more expensive positive one. In events of an undersupplied balancing group, calling the up-regulation always entails extra generation with subsequent fuel and operational costs. Therefore, BRPs' interest is to have predominantly more profitable 'long' oversupply positions in favor of costlier undersupply 'shorts' [62], [84].

3.4   Markets arbitrage opportunity for BRPs

The following two charts (see Figure 3-7) were plotted based on the method provided in [83] to determine the arbitrage opportunities of BRPs between Day-Ahead market and balancing market or any other platform with similar energy products in Germany. This can further hint toward whether it makes sense of trading only on one market or the other. Both charts prove the inefficiency in the arbitrage between two studied markets showing a very weak correlation between the prices. Such a 'poor model fit' could be explained by the delayed publication of the imbalance settlement prices in Germany. Specifically, the reBAP values get known to BRPs only in 20 days after the delivery, thereby hiding the current balance state of the system.

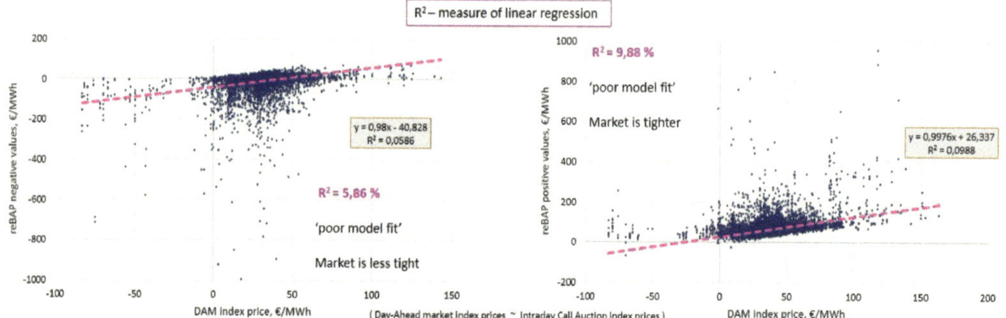

Figure 3-7: *Arbitrage between Day-ahead market and imbalance settlement system in Germany, 2017*

In order to determine the rationality of a BRP in the choice between two markets, the measure of linear regression, R^2 was computed in plotting linear trendline in Microsoft Excel [85]. On both charts it is very low, 5,86% for negative reBAP values and 9,86% for positive. This means the independent variables (prices here) are not well suited to predict dependent variables (type of market). For a 'good model fit' meaning higher predictability in the market type, the linear regression indicator should then usually lie in the range of 60-90%. Noticeably though, the linear regression measure on the right chart with positive reBAP values proves the balancing market at undersupply situations is twice as much tighter to day-ahead as in times of oversupply. In other words, the prices lie closer to each other, not dispersed too much, meaning a narrower imbalance spread and lower incentive for BRPs and hence, slightly stronger correlation of the two markets.

Provided a market has intense price competition leading to the narrowing bid-ask gap, which is likely the case for day-ahead and intraday markets, the market tightness then is high. Taking into account high volatility of reBAP prices and the specificity in the market design where '*net*' and not '*gross*' regulation volume is applied for price averaging [83], these imbalance settlement prices can hardly give a clue on whether this market is tight or not. Apart from this, the two compared markets have a distinct time granularity of traded products, 60 vs. 15 minutes, and the moment of exchange – prior to real-time in DAM vs. ex post in balancing market. As a result, two different markets with uncorrelated market tightness show such a weak correlation.

The insufficiently tight link between the Day-ahead market and real-time imbalance market can be interpreted as a high difficulty to approximate the expected real-time imbalance prices from the forward prices in the day-ahead markets or any other markets [86]. This inhibits the optimal planning horizon for suppliers to schedule their production and consumption and generally boils down to guessing in the dark. So, the only viable and economically efficient strategy left for BRPs is to minimize the amount of deviation and not taking a gamble on the type of market, whereas the latter is officially prohibitive and leads to the loss of the BRP license [50]

By and large, for German BRPs it can mean a nearly impossible way to predict through which method it is more profitable to offset their schedule deviations – either let the reBAP price reveal itself or take action on the intraday market. On the analogy, one can assume that other decentralized options for trading deviations not only encouraged but conversely, economically feasible and could compete with existing solutions. Additionally, alternatives may win in the BRP choice by bringing up rather other aspects that are currently not achievable in today platforms. Essentially, the blockchain-based P2P business models could offer a range of the value-added features as following [87] (Table 3-12.).

Table 3-12: *Blockchain value-added features in the P2P business model*

cost-effective and eased entry requirements	❖	instantaneous settlement
direct access to valuable market data	❖	increased transparency
demand-response-based types of flexible non-energy services	❖	disintermediation

All in all, the above bargains could undeniably supplement the BRP toolbox with rather more leverage options over the imbalances at short notice.

3.5 Germany-wide financial cost from being in imbalance

It is important to size the balancing market share to know how effective the balancing objective in Germany is and whether the policy measures implemented by the regulator BNetzA are in the right direction. The presented cost analysis below serves as an economic basis for potential blockchain options in the area of electricity balancing management.

The balancing absolute cost for the system operator is estimated from the cost of activated control reserves minus the imbalance settlement amounts with BRPs. The vast majority of the balancing costs is made up of capacity reservation component that stands for availability (or holding) of reserve power by PR, SR and TR (approx. with ratio of 10%, 60%, 30% respectively) and it accounted for two thirds or about 70% of the total costs in 2015 [88]. The prices for incremental and decremental secondary reserve capacities are the dominant component that varies due to the rising volume of renewables. According to a series of monitoring reports provided by BNetzA for the years 2015-2017 [42], the pie chart on the Figure 3-8 presents the breakdown of cost structure for energy system services and for network and system security in 2016, while the bar diagram on the right side demonstrates the yearly development of the share of balancing cost in relation to other cost components in *million €*.

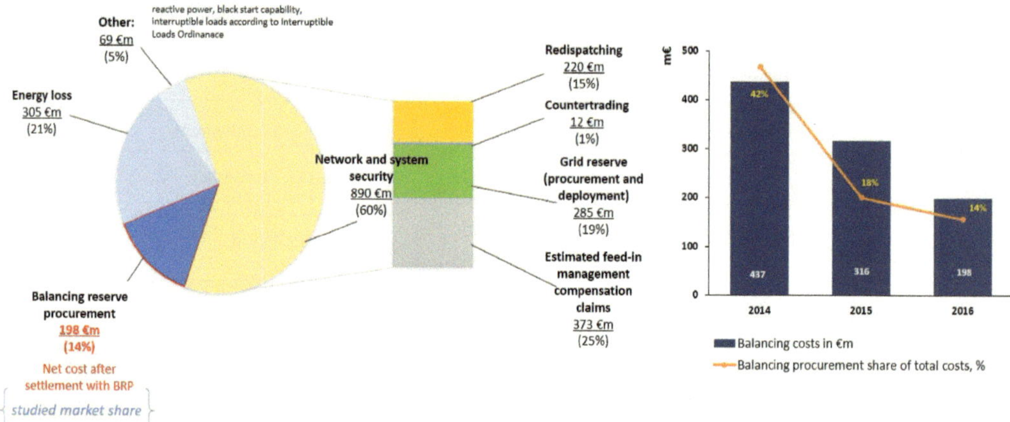

Figure 3-8: *System cost structure and development of the balancing cost share by year*

Crucially, the cost of balancing reserve procurement shows a year-by-year gradually declining character mainly due to the next reasons (Table 3-13) found by studying various literature sources [89], [88]:

Table 3-13: *Measures taken for improving the balancing situation in Germany, since 2012*

1.	Continuously tightening of the rules for BRPs by BNetzA, stricter BRP control
2.	Decrease in technical requirements for control reserves allowing lower capacities, pooling smaller units and short-term contracts
3.	Increase of the spread between imbalance settlement price (reBAP) and day-ahead spot price
4.	Start of hourly intraday trading followed by intraday auction with 15-minute energy products
5.	Introduction of imbalance netting via cross-border cooperation on balancing (*IGCC*) – higher cost-efficiency for the system in order to avoid the counteraction by activating reserves
6.	Shortening of the schedule submission interval before real-time from 1 hour to 15 minutes since past 5 years (up to 14:15, at 14:30 starts intraday phase and DA closes)

The following Figure 3-9 represents a historical development of total utilized control reserves throughout the past six years. The incremental and decremental energy values for net regulated volume (NRV) were downloaded from regelleistung.net for the years 2012-2018. Generally, the maximum activated control capacity declined by 43,4% for positive values and by 69,1% for negative which is the result of above-mentioned reasons and proves their effectiveness. However, from 2017 to 2018 the decrease in activated volumes stagnated that could urge for next policy actions or further strengthening of forecasting quality and portfolio optimization strategies for BRPs. Which may be achieved with promising blockchain technology tools for example. The exact blockchain benefits for BRPs & TSOs and the technology application is explained in the *Chapters 4-5*.

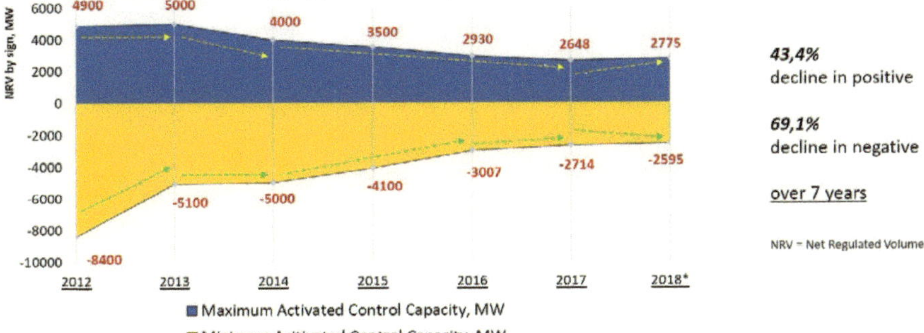

Figure 3-9: *Trend of capacity extremes during the control reserve activation events by year*

Similarly, in [6] it was concluded that for achieving highly efficient system operation that three main policy goals should be pursued by market participants:

- More accurate forecasting combined with improved portfolio management
- A liquid intraday market with atomic granularity of energy products (15 mins or less) as well as moving gate closure to 15 minutes before the delivery
- Increase in transparency, rivalry and liquidity in the balancing market

Next, from the www.transparency.entsoe.eu/balancing/ platform [1] the absolute balancing cost for the system and the BRPs were derived. On the Figure 3-10 the upper curve depicts the total expenses the TSO have after the activation of control reserves (all types: PR, SR and TR), while the lower blue curve, more or less repeating the pattern of the upper orange one, reflects the monetary net settlement BRPs paid to TSO. In the past, it is noticeable that for example Jan 16, May 16 and of recent Dec 18, the synchronized behavior of the cost re-allocation onto the BRP breaks, probably showing either a very good balancing commitment by BRPs and maybe not a strict control, or severe cost from redispatch of EEG-based deviations TSO-borne.

In other words, the lower curve is the income of TSO gained from the paid imbalances caused by BRPs. Hence, the difference or spread between these two curves is the net cost of TSO for bringing the system in the balanced state, while the area below blue curve is only the net cost of BRP. Because the profits or BRP revenues are not mirrored by this graph it is not clear whether BRPs had really been in plus or in minus. The judgement about particular BRPs can only be given based on each commercial profile of an individual BRP group.

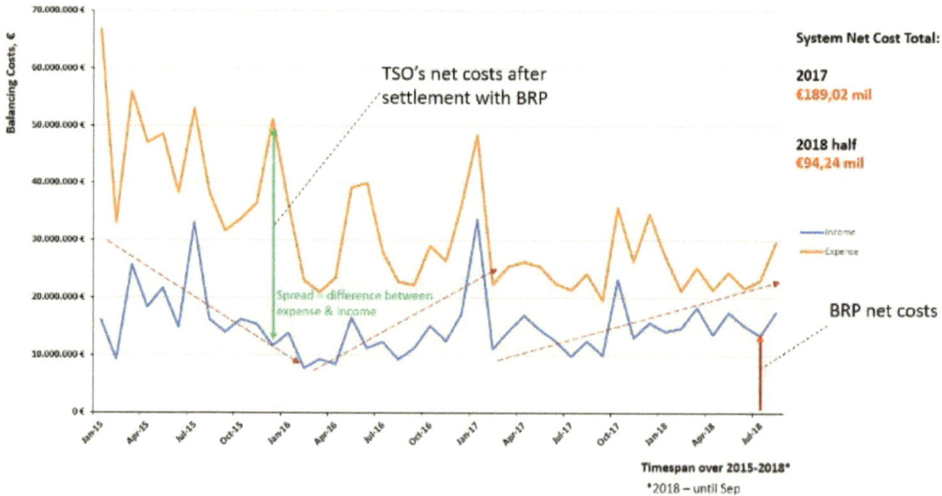

Figure 3-10: *Net balancing costs or total TSO cash flow after settlement with BRPs in 2015-2018*

To get a detailed insight on the financial cost of TSO and BRP separately, the following diagrams (Figure 3-11 and 3-12) show a clear decline of the system balancing net cost and the uneven distribution of the net costs for BRP with noticeable seasonal volatility, correspondingly.

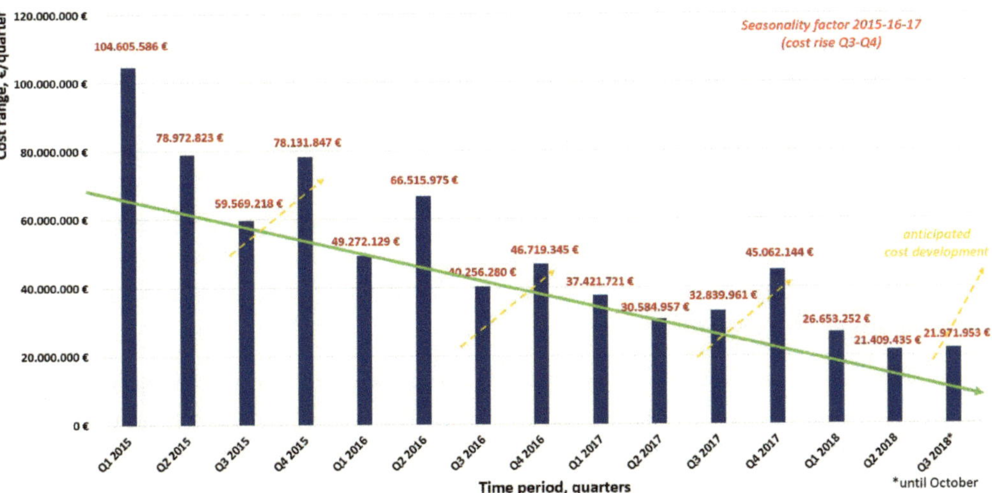

Figure 3-11: *Detailed total system net cost for balancing*

(spread between activation costs and settlement with BRPs)

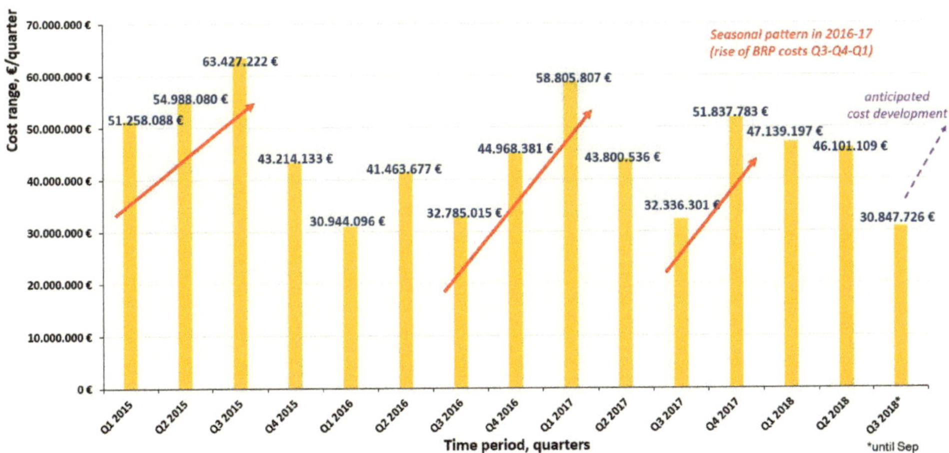

Figure 3-12: *Detailed total BRP expenses charged by TSO for remained deviations*

<u>**Interpretation of findings on the imbalance costs for both BRPs and the system**</u>

Essentially, the trend for total system net cost is descending with big spikes up in January each year explained by the seasonality factor [70], whereas the net cost for BRPs have rather slightly ascending character toward the end of 2018. Initially, by the end of year 2015 both curves fell significantly down, then after stiffening the control over the balancing group by BNetzA [89] the settlement with BRPs started to bring more profits for the system and at the same time became costlier for German BRPs. Noticeably, the spread representing total system net costs for balancing shrank by 70% with a 58% fall of overall control reserve activation cost from 2015 to 2018, whilst the expense component by BRPs has dropped only by 17% which is four times less compared to the net system cost trend. This signals that balancing groups are becoming more and more financially responsible for the schedule deviations and it is still quite costly to maintain balanced positions for BRPs. Therefore, additional balancing tools are proposed in this thesis that are on demand for BRPs in order to avoid the exposure to the expensive regulated market.

Additionally, in terms of the cost component for the utilized renewable energy for balancing by TSO was calculated based on the data downloaded from the netztransparenz.de/EEG [90] platform. The Figure 3-13 represents the relation of the EEG-component in the total system cost structure for 2017 (full) and the first half 2018 to the settled amounts with BRPs that are out of TSO's circle. This bar diagrams show how much (in *million €*) TSO-based balancing groups spent on the balancing by activating only renewable energy sources in comparison to other business BRPs apart from the system.

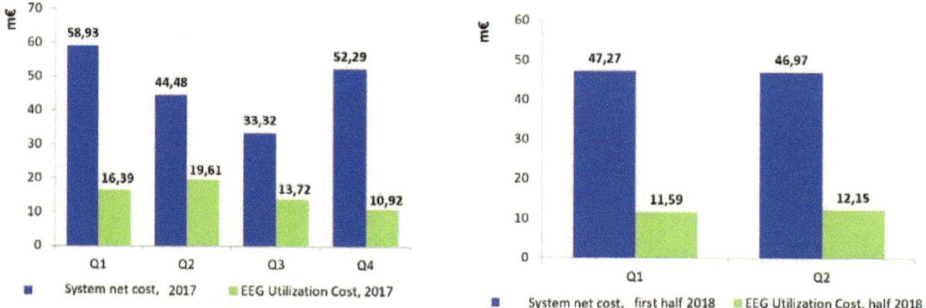

Figure 3-13: *The relation of EEG cost component to the total system net cost on balancing*

Although this cost part is fully covered by TSO and practically there is no financially responsibility by non-system BRPs, it is important to highlight the cost size and belongingness of the balancing RES responsibility. Indirectly, such correlation may imply that had the non-EEG BRP control been stricter, less could have been then spent on the prioritized utilization of RES for stabilizing the system. Also, the studied target class of balancing groups is non-loss, non-difference, non-exchange and non-EEG, and only load & production BRP with physical metering points as discussed already in the *section 2.4.2.*

3.6 Clarification on direction of payment TSO-BRP

Although the secondary reserve had proven decreasing amounts of activated energy (see Figure 3-9) during past years, the financial expenses of BRPs in total were comparatively higher in this situation [4]. It becomes interesting why it turned so. In fact, all data transactions and calculations are recorded over the decentralized data management exchange with market rules MaBiS [65], while the financial money transferring is executed over bank channels agreed in Balancing Group Contract [49]. The Table 3-14 illustrates how the direction of payment between TSO and BRP is determined.

Table 3-14: *Direction of payment between BRP and TSO in cases of oversupply and undersupply*

		Balancing Group Saldo	
		Undersupplied - short	Oversupplied - long
Net Balancing Energy Price	**+** **(short)**	BRP pays to TSO	TSO pays for withdrawn electricity volumes to BRP
	- **(long)**	TSO pays to BRP	BRP pays for withdrawn electricity volumes to TSO

Firstly, the direction depends on, whether the BRP has its balancing group under- or oversupplied at a specific point in time, and secondly, whether the imbalance price (reBAP) is positive or negative. For example, reBAP = -40 €/MWh and BRP holds his balancing group undersupplied at 200 MWh ('short' with positive sign meaning he has lower load value than scheduled by 200 MWh). In other words, BRP's customers in the perimeter consumed less than predicted:

> Forecast schedule = +1000 MWh
> Real-time measured schedule = +800 MWh
> Margin (deviation) = 1000 – 800 = 200 MWh

After matching with the Table 3-14 we see that since the system is long (or reBAP has negative sign), the TSO will pay to the BRP:

> -40 €/MWh * 200 MWh = 8000 €

In this case the BRP receives the financial reward of 8000 € for helping stabilize the system by having the opposite sign of the balance in his portfolio. However, the longer BRP is holding his portfolio undersupplied with negative reBAP, the sooner the payment direction will reverse – BRP pays to TSO [4]. Similarly, the financial penalty will be allocated if the sign of BRP's portfolio balance coincides with system's (reBAP's), meaning the BRP aggravates the balancing of the system and for this he will pay the imbalance tariff. The following bar chart (Figure 3-14) shows how often the system was in short/long as well as how reBAP's sign actually correlates with NRV's sign during the past 4 years.

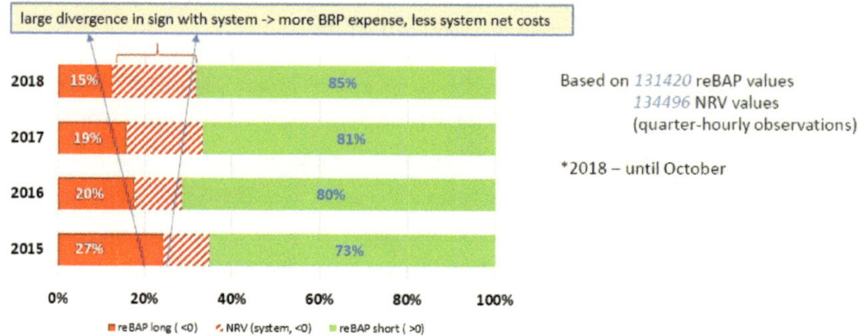

Figure 3-14: *Direction correlation between reBAP and the system (NRV), own observation*

By looking at the reBAP formula (4), one can conclude that if the reBAP sign is the same as of NRV than nominator shall be positive meaning the costs for calling control reserves were higher than revenues (see Table 3-15). In the course of four years the reBAP has had the tendency to become more positive while the share of NRV sign remained stable (approx. 40% with *'minus'* and 60% with *'plus'*), which implies that only nominator has influenced and most logically it is explained by having increasing revenues from activated reserves and less net costs with years. Further reason for small costs is explained by simply lower volumes that were actually activated due to improved forecasting quality of BRPs and enhanced balancing commitment achieved through tightening balancing policies by BNetzA.

Table 3-15: *Probable reBAP sign outcome*

Both nominator and denominator are of same sign results in positive reBAP		
(Costs – Revenues) = negative	**costs < revenues**	$\frac{("-")}{("-")} = +reBAP$
NRV = negative	system was oversupplied	
(Costs – Revenues) = positive	costs > revenues	$\frac{("+")}{("+")} = +reBAP$
NRV = positive	system was undersupplied	
Both nominator and denominator are of opposite sign results in negative reBAP		
(Costs – Revenues) = negative	costs < revenues	$\frac{("-")}{("+")} = -reBAP$
NRV = positive	system was undersupplied	
(Costs – Revenues) = positive	costs > revenues	$\frac{("+")}{("-")} = -reBAP$
NRV = negative	system was oversupplied	

3.7 Risk management of BRP in line with the historic imbalance price analysis

BRP's interest is to always undertake specific risk measures to minimize the probable deviations. By analyzing the imbalance price range, the BRP can see how often the reBAP occurred in a certain price segment. The following results (see Table 3-16) were acquired by analyzing 131420 reBAP values for the period from 2015 to October 2018 that are publicly available at regelleistung.net. Both Figure 3-4 and 3-15 reflect that the imbalance price has skewed historically into positive side, as it can be clearly seen from the Figure 3-14 that the share of negative values has dropped from 27% to 15% over the timespan of four years, while the number of positive values has respectively risen by 12%. The majority of positive reBAP in 2018 took place in the price window 0-50 and 50-100 €/MWh, at 41,94% and 35,92% respectively. Furthermore, the positive prices have moved from 50-100 range into 50-100 by 2018 and doubled in amount in area 200-500 €/MWh. On the negative side, the steady decline from 18,37% to 10,71% in the price range of 0-(-50) €/MWh can be seen over the past years as well as twofold squeeze of -100 to -200 €/MWh from 2,19 to 1% was recorded. In cases of 1000-20000 €/MWh and -1000-(-20000) €/MWh price ranges there was shift into negative part from 0,03% to 0,01% and from 0,03% to 0,05% in 2017 to 2018 respectively. And finally, with *0,01%* extreme reBAP at more than 20000 €/MWh was accounted in 2017 which cause was explained already in the *section 3.2* above.

Table 3-16: *reBAP relative frequency by undersupply (positive) and oversupply (negative), 2015-2018*

** until October*

Price [€/MWh]	reBAP positive				Price [€/MWh]	reBAP negative			
	2015	2016	2017	2018*		2015	2016	2017	2018*
0-50	28,84%	56,82%	45,59%	41,94%	0 – (-50)	18,37%	16,00%	14,75%	10,71%
50-100	35,81%	20,22%	28,77%	35,92%	-50 – (-100)	5,17%	2,94%	4,39%	3,47%
100-200	5,98%	2,92%	4,63%	5,59%	-100 – (-200)	2,19%	0,60%	1,01%	0,77%
200-500	1,78%	0,30%	0,45%	1,18%	-200 – (-500)	1,20%	0,18%	0,28%	0,24%
500-1000	0,20%	0,02%	0,05%	0,11%	-500 – (-1000)	0,24%	0,02%	0,05%	0,03%
1000-20000	0,10%	0,00%	0,03%	0,05%	-1000 – (-20000)	0,15%	0,01%	0,03%	0,01%
> 20 000	0,00%	0,00%	*0,01%*	0,00%	< -20000	0,00%	0,00%	0,00%	0,00%

Having this visual information at hand, particularly the likelihood of reBAP occurrence in either direction and price class, BRPs can estimate the distribution of price changes, reduce severity of loss by knowing what to expect of certain range and see how it develops with years. If the imbalance price is known a BRP can potentially place favorable bids to counteract the system imbalance and thereby enhance their profit-to-loss ratio.

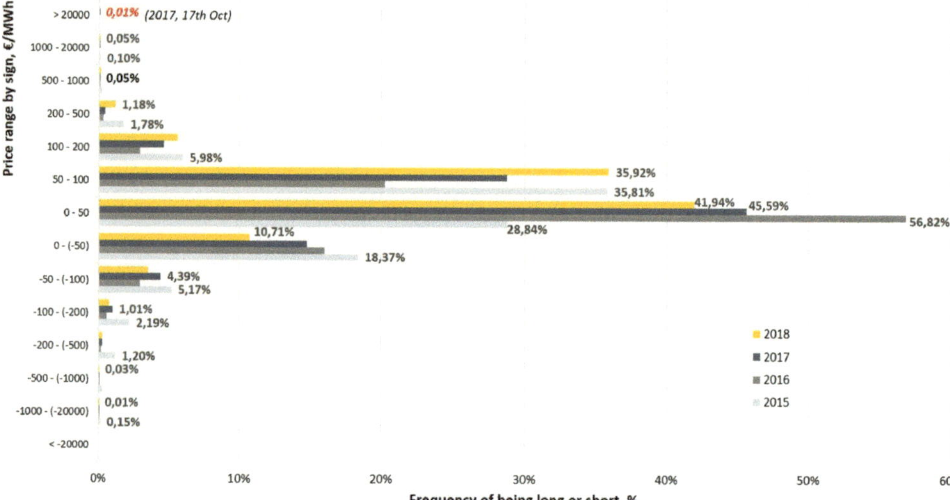

Figure 3-15: *Relative frequency for paid reBAP over 2015-2018 years (*2018 – until October)*

By grouping into categories one can find it useful to find popularity or mode development by studying large datasets. Frequency defines the number of times an event of imbalance price occurs compared to the total number of events within given constraints, e.g. for positive only or for one year. The value assigned to each class is the proportion of the total dataset that belongs in the class. Specifically, based on a sample of 35040 reBAP values in year 2017, we see that in 45,59% cases the imbalance prices happened on the positive scale in range of 0-50 and the proportion of 28,77% of the total reBAP count in the given year in range 50-100 €/MWh. This split into price classes indicates in what direction, short or long, and in what ranges it may potentially dominate the next year.

Nevertheless, relying on the studying historic price development and doing fundamental market analysis does not always help thrive at proper portfolio management because some unexpected events always could derail all 'excellent' strategies. In this vein, ever more critical becomes to apply a variety of financial hedging strategies along with the range of products by energy companies – *'drip feed'*, *'rolling intrinsic'*, *'short-* and *long-term optimization and delta hedging'* are most common to name [91]. To minimize imbalance costs, the hedging can be automatically triggered by price moves, changes in load forecasts or when signals for power plant outages come in. Thus, there is increasing popularity in applying several fintech solutions for portfolio management to grow a successful business with least commercial costs. Blockchain smart contracts certainly can assure same added value of hedging by *if-then* statements (see *section 5.1.4*). But prior to this, it is vital to get acquainted with the regulatory environment in Germany to start using these key instruments with maximum efficiency and conformity.

4 Emerging solutions to avoid high balancing costs

This Chapter takes an unbiased look on the most reasonable method to improve balancing, such as demand response. Profoundly, the current regulatory environment in Germany for the demand side management business of aggregator is reviewed stressing on the conflict of interests with the supplier and its BRP and regulator's disincentive. By focusing on the revitalization of the end-customer in contributing to BRP portfolio balancing, most critical legal requirements in BGM for peer-to-peer interaction over the blockchain platforms are pinpointed. Eventually, bearing in mind the goal of offsetting costly deviations, a service model with a brief breakdown of the main roles is proposed. The model is meant to enable bottom-up marketing of energy-unrelated products connecting all relevant parties in the blockchain network conform to the latest German electricity market regulation.

4.1   Independent Aggregator and Demand Response regulatory barriers

Currently, to escape deviations the BRPs may refer to aggregators who have large pools of flexibility assets managed in so-called virtual power plants (VPP) [92]. Essentially, the aggregators are intermediaries and can act as a community supply provider in Germany (i.e. regional supply) with captured distributed small-scale renewable energy sources, storage facilities and a variety of flexible loads in one large portfolio. By offering software services of matching generation and consumption with integrated demand side management for utilities, marketing of full standardized bid blocks on the power markets is viable.

However, currently in Germany there are still very large barriers for legal utilization of all benefits of demand response (DR) methods. Majorly, the market is blocked for independent aggregators (IA) because BRP-retailer (as identical entity, retailer here equals energy supplier) can interfere with DR [93]. Third party aggregators can affect the BRP-retailer in terms of imbalance issue and the bulk energy issue in cases the IA wants to extract commercial benefits from the utilization of DR. Therefore BRP-retailer may bear the external costs due to aggregator's actions [94]. There is an increasing potential in DR in the energy-only market for BRP-retailer, particularly to minimize imbalance charges, intraday and day-ahead dealing, and arbitrage. However, incentives to balance are not efficient enough for BRPs due to two factors: socialization of the marginal component of balancing costs – reservation of control capacity, and the declining trend of imbalance prices in Germany [45].

Other energy system parties like TSO and DSO also face regulatory barriers. DR leads to higher network charges (§19(2) StromNEV), technical requirements were stated originally for generators and not loads in respect to product definitions, pre-qualification, and finally the regulator incentivizes rather CAPEX over OPEX, meaning that grid reinforcement is prioritized in Germany [95].

It means that there is an urgent need for standardized framework and new business models to enable balancing group adjustments with DR-dispatch. The particular interest is in avoiding intermediary involvement prioritizing peer-to-peer schemes based on a set of rules for applying flexible services within a balancing group. Noticeably, end-consumers are allowed by law to participate in regulation market for down-regulation or up-regulation provision [96], [19]. However, [97] argues that to acquire iMS with 15-minute energy metering resolution to be able to form flex offers is not yet ready.

On the market there are several promising offers with many decades of dealing with DR business models and technicalities. For example – *EnerNOC* [98], having the technical infrastructure for aggregating and marketing at disposal, the company already enables smaller loads to participate and is capable to identify where hidden flexibilities are. Even though such aggregators can take 100% risk for loads pooling, many mandatory agreements have to be collected and signed from various parties in Germany significantly slowing down the entry on the conventional regulated market (see Figure 4-1).

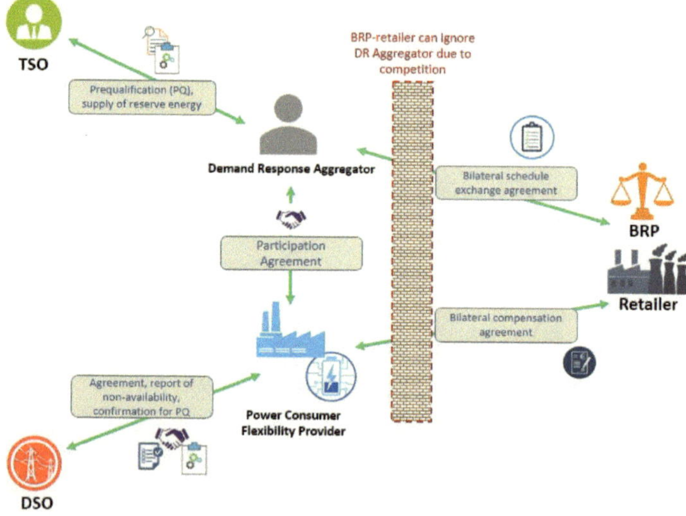

Figure 4-1: *Necessary agreements to acquire for Independent DR Aggregator in Germany* [95], [98]

The following regulatory barriers are critical for flexibility providers to realize DR programs through the standard reserve market in Germany. Wholesale markets like DA and ID are potentially open for electricity BRP-retailer via the implicit and explicit DR. Although retailers are allowed to take advantage of this, the third-party DR aggregators are facing a number of challenges. Although, the DA and ID wholesale markets do not have any explicit legal restriction on the DR activities in Germany, there is simply a lack of requirements in the law and policies for participating in DR. Yet, several other reasons clearly do prevent from actively taking all benefits of the business case for flexibility [95] (Table 4-1).

Table 4-1: *Barriers for Demand Response programs in Germany*

Barriers for DR	Interpretation
Generation oriented	Some markets cannot participate due to prohibitive legislation like for network grid reserve, or based on the product design complexity, e.g. by a capacity reserve
Insufficient incentives	Lacking a framework for the utilizing flexibility procurement as a service, DSO favors costly network upgrades and expansions instead
Pre-qualification requirements	Balancing reserves prioritize rather separate assets with allowed size, and not as collectable parts compatible for pooling
Discounting model for network fees	Favors regular, flat and standardized consumption patterns aiming at maximum utilization rate of current grid infrastructure, but penalizing those who have abrupt leaps and spikes in power consuming, e.g. caused by providing flexibility services [99], [95]. Specifically, as long as large energy-intensive industrial consumers (utilizing more than 100 000 kWh per year) could hold above a specified maximum load hour levels they get discounts according to the charging regime. However, if they want to participate in the down-regulation or provide negative reserve to the TSO with system stabilizing reasons by increasing even more consumption, these discounts will not be valid anymore and they will have to bear significantly higher network charges. Similarly, in case of positive reserve provision, the overall consumption may surpass the lower limit set by the requirement, threatening the system stability
Absence of standardization	The for the role of independent aggregators as a third party in existing DA & ID markets. If they nevertheless want to market their flexible services, the Independent Aggregator must ask for **bilateral permissions** from many parties like retailer BRP – a direct competitor, and DSO, TSO (see Figure 4-1). The major difficulty presents to acquire a bilateral agreement on schedule exchange and compensation payment from consumer's BRP and retailer. The reason for this is lack of interest in collaborating with a third-party aggregator because they would steal their work of offering the same services as supplier-BRP is doing continuously or planning to do so. Therefore, it is insensible to give such an approval to their direct competitor resulting in a large barrier for independent aggregators to enter the market with flexibility offers
A large share of network tariffs	Taxes on top of other fees and levies accounted in the retail prices (50-80%) provide Increasingly low intrinsic market value of electricity, and hence, a very weak price signals of the wholesale market

4.2 The status quo for regulatory environment for the BGM

According to the existing model for BGM, every energy customer is allocated to a balancing group, managed by the contracted BRP being a private legal entity, and its supplier. Also, the power supplier in the role of utility company and the BRP can be the same entity [100], [101].

The metering point operators (*MPO*) outsourced by DSO, manually or remotely read all required certified and authorized energy meters monthly with the purpose to further calculate some components of the energy bill to customers and respective network fees. Once the computation is completed by DSO, he supplies these energy data to the next energy system participants (see Table 4-2).

Table 4-2: *Data flow from DSO after the real energy measurements*

Receiver	Purpose
Power suppliers	accountable for the end-customer billing
TSO	execution of Clearing & Settlement, with actual meter readings at hand he now can compare them with those submitted in schedules by BRPs and compute and bill the resulting costs of activated balancing energy for respective balancing groups who caused the imbalance
BRPs	share received balancing costs from TSO with contracted supplier in order to issue an invoice on their customers respectively

Apparently, for one single electricity delivery for the entire energy market very complex financial transactions processing and metered data exchange must be performed periodically. In principle, every customer of utility companies has to be clearly allocated to a balancing group. Besides, the corresponding BRP is obliged to deposit securities to guarantee that they are solvent affiliate and able cover the distributed costs for the activated balancing energy [49].

4.3 Regulatory challenges for end-customers in BGM over blockchain

With a decentral transaction model based on blockchain technology, the change of BRP role is inevitable. In the new model every energy consumer, producer or prosumer will have to take over the role of BRP. While metering point operators could be no longer responsible for gathering energy values since the record of all energy transactions will be stored on the blockchain [27].

Before blockchain-based P2P energy handling can be legally executed, two types of contract complying with specified rights and obligations must be concluded (see Table 4-3).

Table 4-3: *Regulatory requirements for the end-customer to market its energy products*

Types of contracts according to the trading method		
Peer-to-Peer energy exchange		Marketing on the Reserve Market
Power network access contract	Balancing group contract [49] regulating	Prequalification approval for further partaking in tendering held by TSO
	<ul><li>rights</li><li>obligations</li><li>necessary exchange of info and data</li><li>liability provisions</li><li>rules on the provision of collateral</li><li>rules on termination</li></ul>	Balancing group contract with TSO with mandatory communication of forecast schedules
		Compliance with the rules of the electricity grid access ordinance

This means that participants in the blockchain network will have to comply with abovementioned obligatory guidelines and bear all related responsibilities while exchanging energy quantities no matter by means of what medium, whether a blockchain-based platform, OTC services, via broker or stock exchange. When an end-consumer aims to provide its energy to the balancing power market, there is even more permissions to seize. The exact regulatory barriers for prosumers are discussed in detail further in the section 4.5.

4.4 Adjustments of market roles with blockchain model

The core advantage of blockchain-based transaction model is the capability to transparently allocate the infeed energy values in very short-time lots, up to a few minutes, to a specified customer. In other words, it is highly precise automatized accounting of produced and consumed electricity amounts with dynamic pricing. In fact, the electricity flows physically from a producer directly to the nearest energy consumer. The extensively improved data base allows for more effective and granular control of the networks on the level of distribution and transmission. By simplified clearing process there is significantly lower amount of occasions for receiving balancing energy costs. In the frame of blockchain model, the energy market roles would have the following (Table 4-4) altered contractual conditions [87], [102]:

Table 4-4: *Change in roles in the blockchain-based clearing and settlement processes*

Role	Change
Energy customer	Every energy consumer (prosumer) becomes a Balancing Responsible Party and must therefore fulfill its associated requirements like collateral deposit and risk management. Moreover, the energy consumers now must submit by themselves independently from BRP their load forecasts to the system operator in accordance to all standards of MaBiS
MPO	The MPO will not have to collect meter readings any longer. The customer energy and transaction data are transported online in an automated way, and precisely exchanged over the blockchain technology component, Smart Contracts. To perform necessary computation of grid fees the MPO can access relevant transaction data stored on the blockchain ledger
DSO	DSO will simply extract necessary datasets for the network usage costs calculation from the transactions fixed in the blockchain record
TSO	TSO will no longer need the clearing-related data in the case of fully integrated decentral transaction model. This is achieved with nearly real-time transactions streaming dashboard that occurs only when real energy consumption is validated and indeed the energy amount was delivered and recorded by the energy meter

Furthermore, another regulatory hindrances may cause extra complications ultimately rather refraining from the participation in a peer-to-peer networks [102], [103]:

Financial market regulations

A lot of uncertainty will arise, who will take over the responsibility for the proper functioning of such a system, especially the obligation payments in the frame of the supply contracts. In a perfect scenario there should be an associated responsible authority who meets certain requirements of the financial service provider in Germany. Thus, the next permissions must be acquired (see Table 4-5):

Table 4-5: *The necessary permits to comply with financial regulations in Germany*

I	Compliance with the **German Securities Trading Act** (WpHG) and **German Banking Act** (KWG) to check whether transactions sent over a blockchain are the financial means
II	Application for a license from the **BAFIN** (the German Federal Financial Supervisory Authority)
III	Compliance within the regulations of **REMIT** (Regulation on the wholesale Energy Market Integrity and Transparency) to report transaction data for energy wholesale transactions
IV	Compliance with the rules of financial market regulation **MiFID II** (Markets in Financial Instruments Directive)

Industrial Code

It is not clear yet, what type of energy supplier and whether all energy consumers will have to comply with the **German Trade, Commerce and Industry Regulation Code** (Gewerbeordnung).

Liability issues

There is a need for liability rules to secure the operation of the blockchain-based platforms for such accidents as payment failures, technical defects, data tampering or malicious user activity. The weakest point of failure of existing infrastructure is **power utility companies** and they will be needed to simulate various emergency scenarios in events of a complete or partial system failure.

In general, due to blockchain's relative novelty in energy sector, its versatility and complexity in implementation (e.g. different settings such as public against private blockchain), there is still no uniform legal framework and technological standards on blockchain features. This means that blockchain-based projects will have to operate under the existing regulatory framework in Germany which is almost impossible for a single consumer to comply with in a totally decentralized fashion [104].

4.5 P2P energy trading regulatory view in Germany

As an individual aiming to supply energy for good in a P2P fashion using the existing infrastructure, a prosumer faces lots of constrains and hurdles for realizing such trades legitimately. The Table 4-6 summarizes the energy juridical barriers a prosumer must first overcome to be able to market his or her surplus or deficit of energy to a neighbor [87]. This list is also applicable if this prosumer wants to use blockchain-based marketplace to execute energy-related financial transactions of digital assets. Effectively, the existent environment is stone-resistant to break through and therefore, the support from a third-party service provider like utility company having set all requirements is justifiable.

The question concerning the price sensitivity by a peer-consumer (buyer) is the subject of electric grid tax structure in Germany – in principle, in [26] was proved economically impractical to buy energy with the peer-to-peer method. Taking into account all levies, surcharges, and taxes on top of the negotiated strike price of the peer-vender, the dealt price after tax was estimated about 20% higher than retail electricity price provided by utility tariffs in 2017. So, the prosumer has to be able to offer far less energy prices than those by competitive retailers unless prosumers could avoid using the public grid by trading non-energy units like *rights on switching on/off*. Therefore, this thesis tries to dodge complex and costly regulatory barriers by proposing a secondary market for trading such products (*Chapter 5*), and it is believed that blockchain technology is key to reap all the benefits of shared economy.

Table 4-6: *Legal barriers for prosumer before the P2P trades could be accounted for balancing*

Law	Added tasks and obligations for prosumer
§ 3 Nr. 18 EnWG	Prosumer willing to market its energy regardless of means, is treated as a supplier and hence, should **act as a power utility company** (EVU) with all its responsible obligations and tasks
§ 5 EnWG	Obligation to **register in the (permanent) household supply** asserting personnel, technical and economic capacities to the regulatory authorities
§ 4 StromNZV	Designation as a BRP who is obliged to communicate electricity load forecasts with TSO in line with market rules MaBiS
§ 5 StromStG; §17 (1) StromNEV	**Electricity tax responsibility**. The receiver of energy pays for all taxes, levies and surcharges by consuming every kWh from a given public grid [26] according to the (*Ordinance for German electricity network charges*). Due to the design of

	grid charges in Germany the consumer is bound to pay a complete grid charge irrespectively from the spatial distance between in-feed of electric energy and its draw-off location. This grid charge amplifies when the voltage level is changed by transformers. Thus, if a peer-consumer is not in the same distribution network and is likely from a different balancing group, then it would suffer an enormous grid surcharge accounting for all electrons travelling through many steps of transformation. [105]. By and large, the current regulation for exploiting state-owned physical grid discourages economically the end-customer interaction with a *7,27 € ct/kWh* fee component (a *25% of the 29,44 € ct/kWh* – average el. price for household, 2018 May)
§ 60 EEG:	Power utility company **must pay the EEG-levy** (*Renewable Energy Sources Act*) to the TSO. In 2018 Mai it stood at *6,79 € ct/kWh* [105]. Despite the traded energy was produced and transferred by unusual non-standard characteristics like from a green source or decentrally through a blockchain, the peer-consumer is charged with it to subsidize the incentive of renewables
§ 1 Satz 1 StromNZV	The **contractual design** of the user accessing the network must comply with conditions of a power supply according to §§ 23 ff.
§ 41 EnWG	**Minimum requirement for the contractual arrangement** of energy supply contracts: contract length, price alignment, termination dates, right of withdrawal, performing services, art of payment, liability and compensation for non-compliance, as well as charge-free swift change of suppliers

4.6 Service Model as a solution for P2P energy trading in German regulatory

By virtue of the latest study [106] that emphasizes on the impracticability of balancing group responsibility by a standalone P2P-producer, in this thesis it is studied why prosumer conflicts with its utility company and what is the solution to cope with this problem. Essentially, the generating unit of a consumer, led by a power utility company, must be included in the same utility-associated balancing group that ensures that this generating unit is allocated to this exact balancing group. Based on this, the utility faces a new layout of its balancing group portfolio. Unlike the past, it is no longer sufficient to procure the energy quantities through the wholesale trading based on *only load* forecasts.

Specifically, the consumer can cover their load demand partially owning to P2P generating units effectively disregarding the BRP-supplier's load plans. As a consequence, their BRP increasingly struggles with re-calculating how much exactly it will need to procure via the wholesale market since the metering devices installed at residential customers are not capable of considering household's self-consumption through billing reliant on time frames [107]. As a side effect, the account of balancing group will be unbalanced showing detrimental infeed and offtake quantities leading to the exposure of expensive regulated balancing energy.

Therefore, directing to a *service model* is an inevitable solution to bridge P2P energy trading and the country-specific legal framework [108], [109], [87]. It rules out BRP-supplier's growing complexity of accounting prosumer's self-consumption quantities. A *Service Provider (SP)* connects the members of a blockchain network to each other and to the public grid, thereby stepping out from the role of a sheer commodities provider in favor of the enabler of system integrated P2P handling. The Table 4-7 gives a preliminary breakdown of the SP based on the *USEF* classification [17] and partially on the German *GOFLEX* project [110].

Table 4-7: *Classification of the service provider by rights (own creation)*

Class	In-house Aggregator	Independent Aggregator	Power Utility Company	BGM-as-a-Service	E-Community Operator
Role	Owns all necessary licenses and complies with all the regulatory and contractual requirements stated by the power utility companies	Outsources the needed legal services from a separate power utility company	In German - *Energie-versorgungs-unternehmen* or *Stadtwerke* (end-customer supplier)	Private businesses specializing only in the overtaking the full responsibilities for leading the balancing group management with portfolio optimization tools	A microgrid operator in a geographically restricted area that manages DERs and a bundle of energy consumers

Having the full package of BRP's legal privileges at disposal, each class of service provider must register a single balancing group per one blockchain network and provide their end-customers with a technical blockchain application which data processing regulatory-conform is. All customers conclude a physical or electronic-based contract with a SP and are allocated to their blockchain networks correspondingly. The peer-to-peer transactions can be now accounted for the official scheduling process with TSO inherent to BRPs in Germany. Next, it is important to differentiate among several types of the blockchain network members (see Table 4-8.).

Table 4-8: *Divide of the customer roles in the proposed service model*

	Types of blockchain network members in the service model
Only consumer side	- Household without generation unit (SLP) - Industrial consumer (RLM); shiftable and disconnectable load
Producer side	- Small-scale power plants, known as distributed energy resources (*DER*), e.g. a private PV system; remotely controlled and disconnectable

Prosumer side	- Household with renewable generation units
	    • prosumer offering its electricity by direct subsidized marketing [107]
	    • prosumer not marketing its produced electricity using premium system and thus managed as sub-balancing groups must abandon a right of market premium [111].
	- Storage facilities (batteries, controlled fully remotely)

Based on [112] a considered load flexible model in this thesis presents the ability to move the consumption either backward or forward because they cannot store electricity with a rise in the consumptions profile. This opts out the buffering capability (which is inherent to thermal inertia), and hence can be split into two parts: first, disconnectable loads, which consumption can be interrupted losing the ability to recover; second, shiftable loads – can be rescheduled not involving any change in the consuming pattern.

The next Chapter details how blockchain technology can be applied in relevant steps of BGM from the conceptual and organizational perspective with a proposal of a business case for end-customer exchange of flex services. Also, from technological standpoint, the introduction of blockchain is found a perfect fit to enhance efficiency of both BRP and TSO in the established framework of market rules MaBiS.

5 Incorporation of blockchain components in the BGM

This Chapter unfolds possible areas of the blockchain technology embodiment in the balancing group processes by breaking it up into two parts. Firstly, the emulation of a blockchain-based end-customer-centric decentral marketplace is proposed. In such a market, consumers can exchange abstract non-kWh units among themselves in respect to the service model conform to the German regulatory described in the previous *Chapter 4*. Following this, the service providers interact with each other to compensate remaining deviations over the blockchain-featured *Bridging Protocol*. Secondly, the feasibility of blockchain application in the clearing and settlement processes within the MaBiS format is studied. The transactions there present transferring of energy-related dynamic and static values, as well as of financial assets. Ultimately, a blockchain use case is revealed for the internal sub-BGM lodging in cascades.

Up to now there has been a large uncertainty and research [109] on whether P2P trades are viable to perform using the blockchain technology in the German framework for BGM complying with the standardized market rules. Trading in today energy system has many steps involved, to visualize the complexity of the customer-to-customer energy handling mechanism that reflects peer-to-peer relationship the following Figure 5-1 was adopted from the-state-of-the-art study [87]:

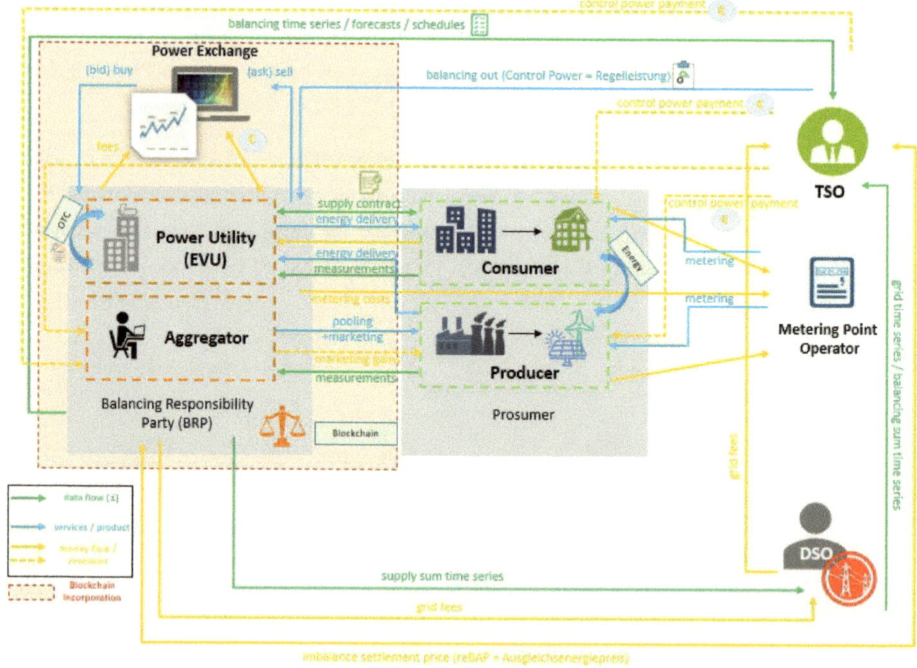

Figure 5-1: *Reasonable incorporation area of the blockchain technology in the existing energy trading mechanism in Germany*

5.1 Secondary balancing market concept to facilitate decentralized balancing

Focusing on prosumers, there is still a host of regulatory barriers to realize absolutely decentralized peer-to-peer transactions of various energy products and balancing services over the unregulated blockchain applications. Therefore, the proposed service model was developed as a solution to overcome the legislation stumbling blocks and enable the bottom-up movement. In this regard, a schematic simulation of the interactions among market parties is presented on the Figure 5-2 and will be explained in detail within the sections below.

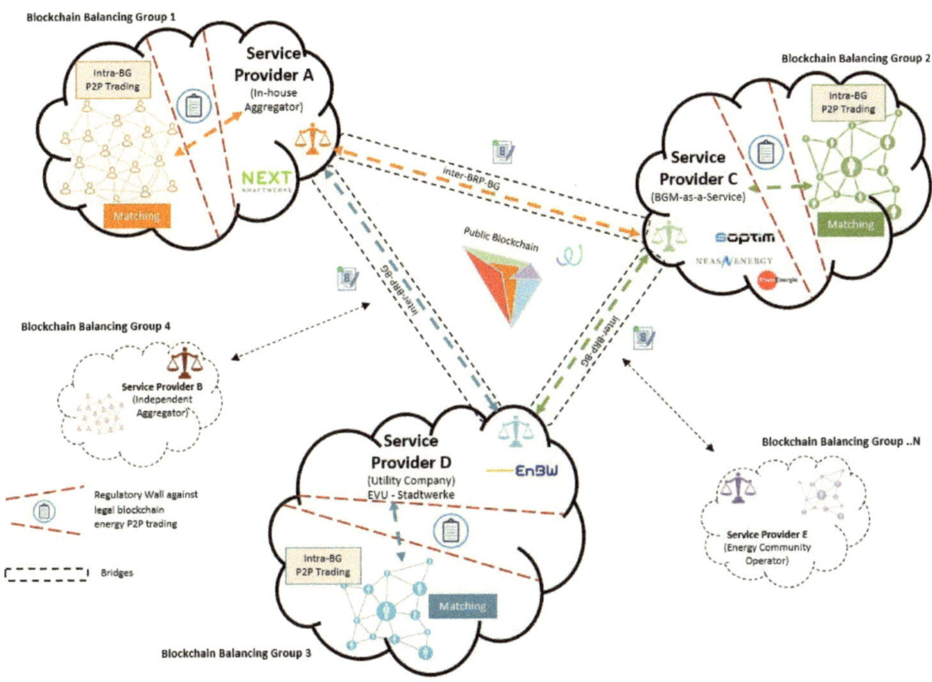

Figure 5-2: *Schematic simulation for the intra-Group, in-Group and cross-BRP assets exchange in line with the proposed service model*

For the technological realization of the proposed scheme, namely the architectural intricacies in choosing and implementing right blockchain components, the master thesis [11] is recommended since it highlights the most suitable generic architectural elements for the energy management of multiple parties in decentralized P2P networks. Some concepts were partially adopted into the current thesis.

5.1.1 Balancing scenarios split

The schematic division of BGM by blockchain networks has the following structure. Each balancing group (BG) has one Service Provider (SP) who must either possess itself or outsource all necessary privileges, licenses and rights for administrating a balancing group. In other words, this group operator has to be entitled Balancing Responsible Party. In each blockchain-BG (BBG) there is a network of customers who are in legally contractual relationship with its SP.

Every P2P network has its local blockchain-based market with matching mechanism of *'quotas'*. The versatility of customer preferences urged for the product differentiation setup which best expresses the utility of consumers and exchange reasonability in P2P models [113]. Moreover, the efficiency and ease of the implementation of a so-called *MBED-model* that realizes multi-bilateral trading of such products in truly decentralized peer-to-peer market settings proved by [114] confirms a non-blockchain technical implementation. Prospectively, it is interesting to see how this model can be applied on the blockchain.

In this thesis it is thought that the price for quota can vary due to the change in desirable properties a customer may opt for. Essentially, one quota equals one mega watt of the day-head schedule submitted by a load-customer. The quotas are not identical, they can have the following attributes (see Table 5-1):

Table 5-1: *Customer sensitivity due to product attributes*

Properties of quotas		What will be seen?
availability of storing energy or adjusting load	❖	yes/no
sign	❖	negative/positive
type and origin of the notified electricity amounts	❖	wind, solar, gas, etc. green/conventional
distance	❖	e.g. 20 km
social aspects like a rating for successfully provided services	❖	e.g. 5 stars
allocated under specific BRP	❖	e.g. BRP A (with EIC X)

Generally, quotas may be interpreted as schedule deviations. So, the local blockchain-based market allows for exchanging these quotas in a dynamic manner on the customer-to-customer level within one BG. The pricing boundaries are reliant on the signals the SP sends onto the platform. The logic for determining *right* price signals is provided in the section 5.1.4.

These local deals in order to be eligible to MaBiS processes they have to conform with a long list of regulatory requirements and bureaucratic hurdles. This task is managed by the SP of a respective BG cloud. The SP for BBG assets can take up to 5 roles in today setting of German energy businesses, the Table 5-2 details activities occurring by each type and gives examples of such operators.

Table 5-2: *Detailed interpretation on the roles in the setting of proposed service model*

A	As an **"In-house Aggregator"** that accumulates a large flexibility pool consisting of DR assets provided by industries, businesses, and residential end-customers. The resulted pool is converted into several products to meet the needs of a range of stakeholders. It associates with its own retailer-BRP that has all necessary tools at hand for dealing with portfolio optimization, marketing and complying with local regulations. In other words, an integrated aggregator runs its business without any involvement of a supplier or any other market party. This role plays the German largest flexibility pool provider *Next Kraftwerke GmbH* [115].
B	As an **"Independent Aggregator"** – it is not identical to the supplier and to its BRP, though similar to the *In-house Aggregator* from technical equipment perspective but lacking all necessary regulation permits. Therefore, third-party aggregator must externally contract with the potential competetor retailer-BRP and collect lots of bilateral agreements as discussed in section 4.1.
C	As a **"BGM-as-a-Service"** – mainly has state-of-the-art weather forecast modelling either infeed from a reliable source, then fundamental and technical analysis for the market price developments coupled with real-time price data supply. They also have a gateway to TSO and know how to do all necessary administrative scheduling processes over MaBiS. For example, to name a few - *Neas Energy GmbH* [116], *Rhein Energy AG* [117], *SOPTIM AG* [118] – are leading companies in energy management and revenue optimization services for independently owned assets can take over these responsibilities.
D	As a **"Power Utility Company** (known as *Energieversorgungsunternehmen (EVU)* with *Stadtwerke* in Germany). They either perform all relevant data exchange and forecasting by themselves, which is normally the case, or outsource this job to before mentioned *"BGM-as-a-Service"* company. As a rule, municipal utilities have very large client bases of consumers. German utility companies, such as *Stadtwerke Schwäbisch Hall GmbH* [119], or *EnBW Energie Baden-Württemberg AG* [120], etc. are very well equipped with all necessary tools for leading BGM. A good example is the project *Tal.Markt* [121] blockchain platform served by the municipal energy provider *WSW*. It is already creating local value and new streams of revenues by operating as a matchmaker and coordinator between the local producers and dwellers allowing to choose their own energy mix in the Wuppertal area.
E	As an **"Energy Community Operator"**, a manager of a local *energy collective* [122], [113] who has access to the energy markets and is capable of aggregating variable consumers, prosumers and RES-producers, small commercial companies and industrial prosumers, residential sector as well as electric vehicle charging systems. The microgrid manager may act also as a classical supply company optimizing the supply and demand in home networks with self-consumption. Some companies, e.g. *Lumenaza GmbH* [123], with blockchain-based settlement model already provide intelligent energy management solutions for utilities (e.g. *SWW Wunsiedel*). As a BRP it enables regional balancing of BGs by directly connecting end-customers in the P2P network.

5.1.2 In-Group balancing

The P2P trading in such a balancing group is performed on the blockchain network that only accessible to the BRP and its customers of this balancing group. At the moment, only a single P2P network is considered where its SP oversees the trading inside of it (see Figure 5-3). While in the sub-section 5.2.1 an active interaction between end-sub-BGs supervised by a Link-BG manager on different bottom-up levels over a blockchain medium is proposed. In particular, the sub-balancing groups are lodged in the cascades where one layer is nested inside of the other downstream off the financially standing BRP in the head. Unsurprisingly, this portfolio organization strongly resembles a Russian doll technique – *matryoshka*. By combining this approach with blockchain already a handful of P2P energy trading projects intensively research and promote the uniqueness and efficiency of balancing the grid in such an edge-up style [124], [125].

Figure 5-3: *Intra-BG P2P matching over the blockchain*

The assumed concepts for P2P transactions are customized permissioned blockchain with the Proof of Authority consensus mechanism, P2P communication protocol, a decentralized file storage and state channels according to [11]. The choice fell on this due to the reasons of reduced economic costs, latency and retaining privacy that achieved in the compact networks that want higher performance.

The communication of participants equipped with *intelligent measuring systems* (*iMSys*) forming a network is realized over the peer-to-peer messaging protocol and lightweight blockchain client and decentralized application parts. Not only remote energy monitoring and control with dynamic tariffing functionality is key for iMSys, but also it can connect to a network, host light blockchain protocols, has enough storage and computational power to process and make use of the cryptographic hash functions. A number of peers with *authority* status is selected by the community based on the technical capabilities for adequately hosting blockchain server full nodes. This is because it requires to perform difficult tasks of providing consensus, hosting protocols and then interexchange with other BBGs. Therefore, there is a prerequisite that the SP must maintain an authority node in order to inter-operate with other BRPs' blockchains.

The proposed scenario for P2P trading does not imply the actual physical transfer of energy units, kWh or such. Instead, the users of blockchain-based secondary market are granted with the opportunity to offer their willingness or capabilities to 'turn on' their energy assets at times when it is needed. At every customer location the market agents can provide the availability of soaking up a specified demand or ramping up in capacity in the times of BRP imbalances. The proposed units – *quota prime* – can serve as a trading digital mediator. These units reflect directed deviations with attributes of availability, liability rank, and usage rights that cannot be taxed by the regulator or at least at a mild effect.

The iMSys record and analyze all energy-related values every 5 minutes at an energy asset location. It can be dynamic timestamped consumption/generation values and static data such as location name, energy source type, period in exploitation, some owner data, geographical situation – distance to other nodes. The schematic life cycle of the quota-based secondary market is illustrated on the Figure 5-4.

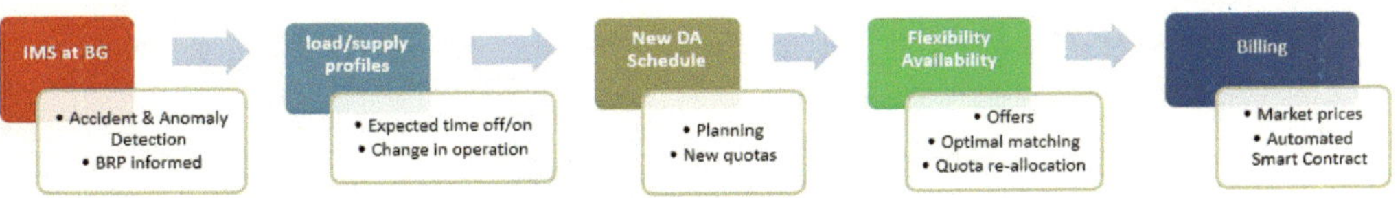

Figure 5-4: *Life cycle of the quota-based secondary market*

After aggregating real-time consumption data, based on historical load profiles day-ahead schedules are issued automatically by iMSys. Initially, every 15-min MW-energy value is transformed into *quotas*, which by common sense should provide a limit for a certain amount (similar to *binding quota scenario for congestion management* by *Agora* [3]). The limit is the maximum notified amount of energy for the next day. For instance, a schedule was created with 40 MW meaning 40 quotas are given until delivery. While for the household-level customers, a fraction of a quota is applied, e.g. 23 kW equals 0.023 quota.

Whenever an anomaly or any irrational alteration on consumption/generation is detected by iMSys the quota amount is re-calculated, effectively showing individual's deviations. For example, one production line of a commercial or industry customer got compromised, or simply unexpected machinery failure occurred. It implies a sudden drop in the consumption profile and potentially a financial loss from future allocated imbalance costs with reBAP. By observing and reading this change, iMSys comprehensively optimizes and updates the day-ahead quotas into *quotas prime*. New quotas prime may have two signs positive or negative. Positive means undersupply and negative oversupply.

Peers can exchange their quotas prime based on a price range provided by their SP. The iMSys decides on the best economical deal for BRP based on the data retrieval from wholesale markets and imbalance market (prediction model). The key aspect for BRP is to not correlate with the system imbalance in a sense of not aggravating it and thereby avoid painful imbalance penalties as discussed in the *Chapter 3*. The incentive mechanism against market prices is discussed further in detail in the *sub-section 5.1.4*.

5.1.3   An emulation instance for the quota market model

Blockchain-adapted and reliable iMSys devices post short-term bids at the energy asset location on the BRP-hosted platform. The built-in options enable increased hedging and reliability capabilities.

Customer nodes are trying to replicate BRP's trading by means of a market that imitates real prices of reBAP and DAM but with blockchain-added attributes. The BRP is always reliant on the flexibilities, system state models and other real-time offers along with weather and prices forecasts. In order for the customer to use the utility of such tools at hand, the BRP sends price signals to simplify network users' rationale. After a base P2P trade took place, the flexibility is activated in the portfolio. This in essence optimizes its state leading to neutral.

As a rule, the higher the reBAP, the more motivated traders are. Since the reBAP trend has a historically declining character, conditional markup was added to keep the penalties high. The BRP aims at most attractive price zone for dealings held by portfolio customers. However, due to intransparent internal management, root customers do not know the state of BRP portfolio to be able to help him. Therefore, BRP provides these metrics for the orientation and competition chance.

The Table 5-3 provides a basic view on the assumed market rules for the proposed quota-based secondary market mechanism built with blockchain functions.

Table 5-3: *Interpretation of the quota model conditions*

Quota model	End-customers are faced with a choice, either *standby* or *earn* by turning on, not entailing the imbalance tariff charges. The products with merit-based markups can be exchanged only if permission was granted after reaching a set threshold programmed by smart contracts
Customer perspective	The load buyer will agree on prices as long as his operation is profitable, while in case of generators LCOE must be lower than the offered price range signal. So, liability and responsiveness ranking assigns discounts on purchaser quotas.
	The seller looks for higher prices or an actual ability to offer their services. So, properties of vender services are adding value to their quotas, thereby increasing the price
Common sense	The participant who prefers to consume more in excess will sell his/her margin because of fixed fuel, variable and operational costs - need of ensured return on

not idle asset utilization. The price fairness is reached by the free access to market prices with the competing markup feature.

Market manipulation is excluded by *Proof of Price* (or *Proof of Value* in activation)

Lets assume a market setup of only three loads which experience undesirable events during their operation leading to the changes in profiles notified in schedules (see Table 5-4, Figure 5-5).

Table 5-4: *Market-specific parameters for intra-balancing-group trading*

	Type	Anomaly spotted	Expected time off/on	Schedule DA	S' after box reading
Load 1A	RLM, controllable	1 line off	4 hours off	10 MW	8,33
Load 2A	RLM, disconnectable	machinery repairs	16 hours off	30 MW	17,2
Load 3A	RLM, shiftable	willing to work more	on	60 MW	60

	Flex availability	q	q'	Strike price	Result
Load 1A	+ 1,67 MW	10	1,7	Script	Neutrality
Load 2A	+ 12,8 MW	30	12,8	Script	Neutrality
Load 3A	0	60	-14,5	Script	Gains

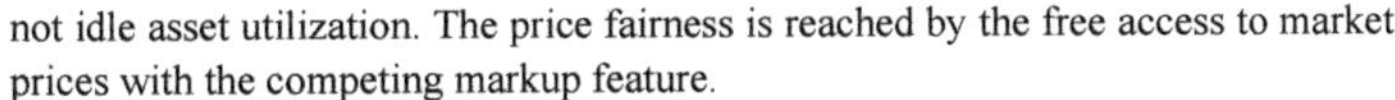

<u>Assumptions:</u>

- q – *minimum threshold (fixed) for activating energy assets, allows to buy, appreciates response*
- q' – *activation rights* that are private goods comprising permits for feeding in and offtaking in a designated BBG guarantee deviation-free portfolio management (adopted for BGM from [3])
- q' = binding quota prime reflecting *deviations* after DA scheduling in real-time

  if $(0 < q' < q)$ – only sell ; if $(q' < 0)$ – can buy, higher preference to activate energy assets
- $q = 0$ – no deviation (before clearing) rule of allocation applies – sets $q' = 0$
  - gets a total sum of deviations with the opposite sign : > $q_{3A}' = - |q_{1A}' + q_{2A}'|$
- Each iMSys agent has goal of neutrality where $q' = 0$ after trading with other agents
- Script ran by iMSys solves BRP cost minimization problem (if not risk-averse)
  (if desired – profit maximization problem based on aggressiveness constraints [126])

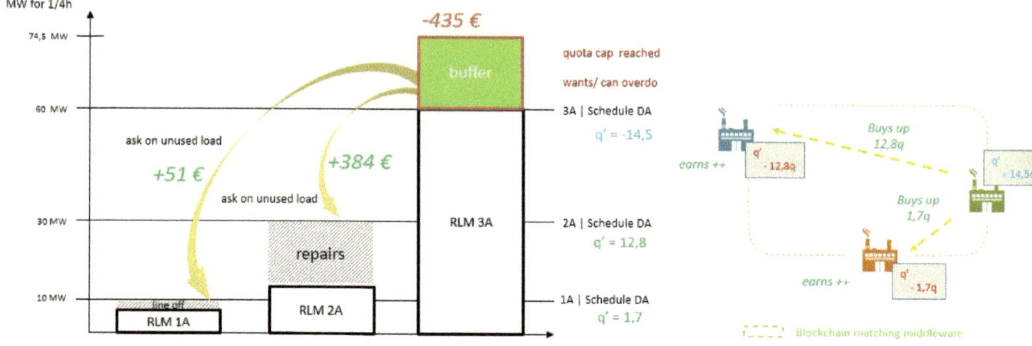

Figure 5-5: *Emulation of an intra-Group scenario for P2P inter-load-only trading*

In the example above three different large consumers were chosen. Load 1A, Load 2A, Load 3A submitted day-ahead schedules of 10 MW, 30 MW, 60 MW respectively. Schedules are managed by their SP. At some point in real time in the phase of schedule correction after day-ahead GCT it became clear that planned consumption could be not fulfilled by two customers Load 1A and Load 2A. The controlling BRP received signals from iMSys alarming that two loads are off and unable to deliver.

> *Note: an intention to arbitrage and gambling on the system imbalance direction is excluded. The main goal of BRP is to be in a neutral state, net deviation should be as close to zero as possible. This implies that BRP customers aim at zero sum as well.*

The smart box executing the algorithm written in the script (a simple logic provided in *sub-section 5.1.4*) defines the striking price range where the loads can bid. The price assumed in the example is *30 €/MWh*. Based on the estimated *time off* by each load the iMSys computes new schedules and subsequently alters base quotas into new *quota primes*. Only owners of quota primes are allowed to trade their deviations among themselves in a specific price range to avoid correlation with the system imbalance.

In the outlined scenario the largest load (3A) wants to produce more, even though it has reached already the limit cap of 60 MW. The iMSys detects its willingness and offers earnings to its peers who went on repairs and failed to deliver as planned. Once the confirmation was received, the Load 3A is granted with quota primes having opposite sign which amount is equal to the total deviation in the group. This enables Load 3A to bid within a price range provided by the script. Considering offer's attributes peers agree on price and effectively offset their undersupplied quantities, whereas the largest Load 3A produced more goods and was not financially penalized for its surplus with unfavorable reBAP.

In this scenario the managing SP benefits from the DR-aware consumer who was able to counteract under an undesirable turn of events. The responsiveness is also considered to be topped with premium by a ranking system. The trading is performed on a blockchain platform, which implies disintermediation and eased access to market differentiated products. With real-time pricing algorithm the prices are linked to day-ahead (or intraday) wholesale market prices read from EEX or EPEX Spot SE. This gives an access-free opportunity of offering flex services at the true market value avoiding high price volatility, large fees, all necessary permissions, and entry requirement hurdles [13], [87].

Blockchain-wise, a quota can be represented as a unit for exchange like a *stablecoin* [127]. A stablecoin is a class of cryptocurrency that functions as a store of monetary value and used as a unit to reflect the account balance. Fundamentally, stablescoins are backed by reserve assets (as a collateral) and their demand and supply are also timely controlled by authorities to prevent any drastic changes in valuations. Thus, when coupled to a public cryptocurrency-to-fiat exchange, it is usually volatility-free in its valuation guaranteeing relatively the same value over the entire lifetime. It allows to link traded quota primes with real monetary value like $€$, such that the customers could estimate the trade-off.

The Table 5-5 shows the most important drivers for P2P trading by means of blockchain.

Table 5-5: *Five key traits of P2P dealing in the blockchain balancing group*

Consumer-centric	Censorship-resistant	Bottom-up rights	Responsiveness	Rewards
Activates the awareness of silenced roots	No one can stop from bidding, all on equal terms, immutable, irreversible transactions	Exchange the capability of availability	Supporting in critical times may lead to substantial savings	Eagerness to earn extra by helping others

Once the internal customer-to-customer deviation trade took place, now their SP can account these deals for correction of his overall portfolio imbalance. Even if after P2P event there is some imbalances in BRP portfolio left due to unfavorable circumstances, it has another option to inter-exchange them with other SPs who have the opposite sign and matching values in their portfolios. This exchange can also take place via blockchain through specific bridging protocols that allow multi-chain interaction.

5.1.4   Inter-Group or cross-BRP balancing

After intra-group trading there is a further possibility for SPs to adjust their schedules not directing to the traditional methods such as OTC or Spot markets. This may be managed through *inter-chain communication bridges over a public blockchain* (see Figure 5-6). A BRP-SP of one blockchain network in this scenario wants to send its total deviation in the form of digital assets (via tokens) from one BBG to another. One token is designed as a stablecoin equal to 1 MWh of schedule for example.

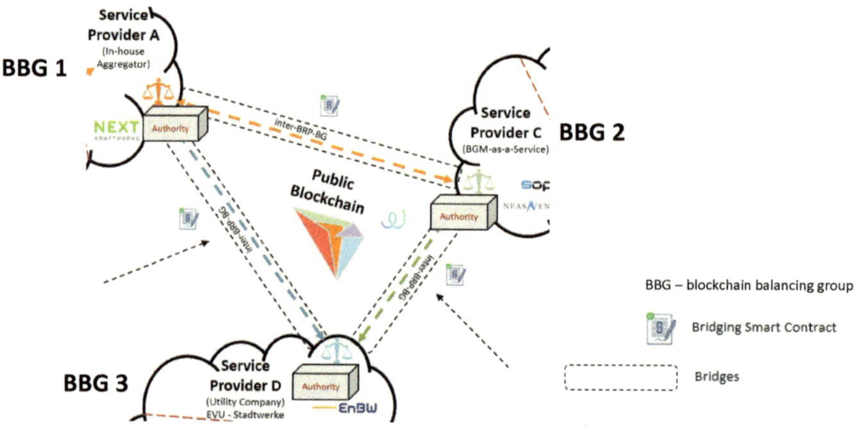

Figure 5-6: *Inter-BRP business dealing*

In the scenario with multiple blockchains there is a difficulty of interchanging the content because consensus and validation mechanisms are located in each individual chains [11]. For this cross-chain exchange there are already many emerging projects in place. Bridges by *Parity* [128] and *Polkadot* [129] are two good projects aiming to solve this problem by developing a *bridging protocol*. Although there are other promising developing projects like *Cosmos* [130], *Plasma* [131], and *MultiChain* [132] with a similar goal to track tokens in connected blockchains with the ability of exchanging them directly, the current section does not compare nor go deeper in their provided solutions but explains the main idea.

Parity's protocol features *ERC20 token* interexchange in line with deposition and withdrawal of cryptocurrency on/from a *Bridging Smart Contract*. The unique aspect proposed by Polkadot technology is *bridges*, a link between blockchains with independent consensus. Polkadot is promised to be a heterogeneous multi-chain technology that promotes the anonymous and secure exchange of data and payments between different blockchains. Yet it is still under intensive development stage.

Heterogeneous chains can be explained essentially by two things. Firstly, networks have different layers, in other words they can differ in their *networking*, *consensus* and *application* implementation parts. Secondly, a blockchain is basically sovereign with its own set of validators who maintain the chain and agree on the next block to commit to the blockchain (e.g. miners if PoW). And it is important to be sovereign since the validators are responsible for modifying the state. Here, they are majorly SPs.

The SPs as well as other end-customers who installed with authority nodes can communicate the total deviations over the public blockchain such as the *Tobalaba* platform hosted by *Energy Web Foundation* [36]. The Tobalaba platform is poised to be the standard platform for energy-related use cases in the world that is why it is considered by this thesis. The infrastructure design of a public blockchain considers a more persistent sustainable ledger with increased transparency and immutability owning to a very high amount of hosting nodes compared to fewer in permissioned ones.

A public blockchain thus allows to backup hashes, bidirectionally inter-operate with other permissioned blockchains hosted by SPs as well as transferring large amounts of financial assets using local tokens. With hashed data backups validated authorities like SPs and fairly elected end-customers, can post immutable entries of deviations that will be visible to all authority nodes on the public blockchain and further used for inter-trading. Lets say, from an industry customer to the SP, then to another SP who is missing this certain amount to optimize its portfolio. The economical rationality for SP-to-SP trading is touched in the next sub-section 5.1.4.

Managing the deviations via the intra-group mechanism cannot be always sufficient because there might be some unfavorable situations when there is no possibility of offsetting the imbalances. For example, when all peers are undersupplied and the system is short or vice versa. As well as, the situations when no one is willing or capable of providing a flexibility service within a balancing group due to a lack of resources. Similarly, alternatives from other SPs may seem more cost-effective providing a groundbreaking set of values alongside compared to traditional pricey exchanges with minimalistic product details. Therefore, there is a need for this trustless inter-BG communication. This can be achieved and featured by bridges that are not necessarily part of the network and enable interoperability among the BBGs preserving reasonable security standards. The cross-chain asset exchange and portability designed as a cross-BRP alliance allows merging separate *'BG-islands'* to realize the ubiquitous value of interconnection, trading and ridding of deviations.

5.1.5 Incentive mechanism against market prices

The following algorithm was logically developed for blockchain secondary market (Figure 5-7, 5-8). The assumption that was made is that imbalance price is known in real-time which would allow a more weighted choice for BRP-SPs to make when both sending price signals to their customers or inter-exchange with other BRP-SPs. Relying on the price signal, market agents by trading with each other peer-to-peer should not aggravate the balance of BRP portfolio and vice versa always keep it in a favorable state leading to the neutral state which is the main goal of the German BRPs [76].

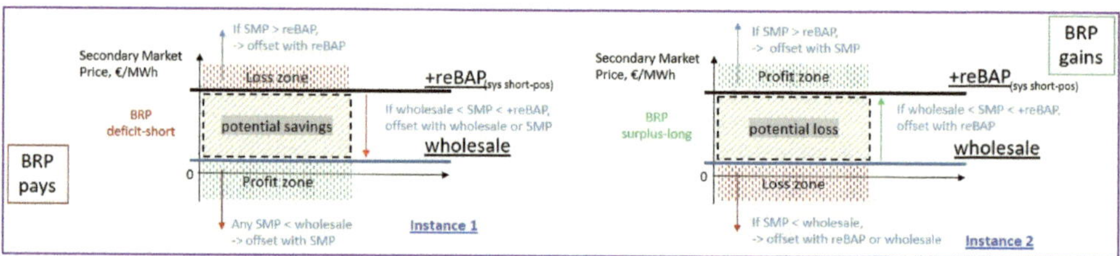

Figure 5-7: *Script logic for sending the right price signal at the state of system short, positive reBAP*

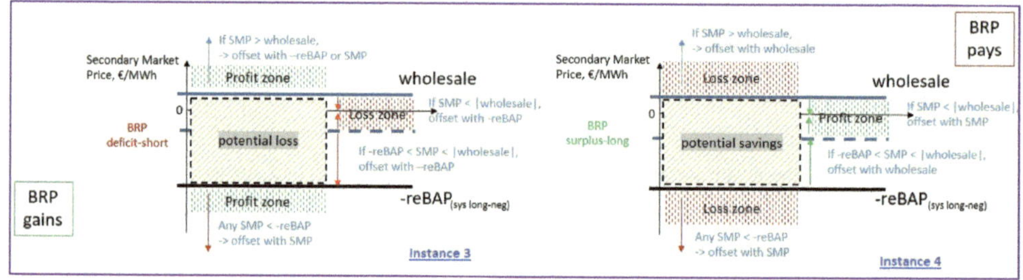

Figure 5-8: *Script logic for sending the right price signal at the state of system long, negative reBAP*

In the graphs above the acronym *SMP* stands for the *Secondary Market Price*. It competes alongside with imbalance price (reBAP) and wholesale prices. Essentially, the emulated trading logic relates positive and negative deviations represented by *quotas prime* in a P2P network (after rescheduling according to the discussed concept in *sections 5.1.2-5.1.4*). The flexible energy assets' capabilities that are willing to offset deviations and were approved by the market rules will be utilized depending on the available price range recommended by their Service Provider with the status of BRP (see Table 5-6).

The limits of SMP are determined subsequently by expected imbalance cost, which is reBAP times deviation volume in each 15-minute time slot, and the wholesale market price. For the wholesale market, the Day-Ahead market price or Intraday Call Auction price at EPEX Spot is recommended as *a reference price*. Likewise, the reBAP values should either be modelled or fed in from official sources, provided that the nearly live posting of imbalance prices was enforced and hence made publicly available by the regulator BNetzA (e.g. Belgium and Netherlands publish nearly real-time). Both prices may serve as a lower or upper border as a function of the direction of net deviation, either shortage or surplus.

Table 5-6: *Clarification on decision making by BRP for sending cost-effective price signals*

Instance 1:	Whenever the system is short meaning an excess of energy, but BRP has deficit in his portfolio than the upper boundary is imbalance price and the trader would strive to buy lacking energy at a price lower than this reBAP. This is simply explained by the objective of the deviation cost minimization with a least possible price. Thus, he would give a price signal in range lower than this upper limit leading to potential savings. *(Quadrant 1)*
Instance 2:	In case the BRP has a surplus in his portfolio at the same state of the system (short), the BRP should encourage his customers to sell at the prices higher than the reBAP because what lies lower is unprofitable and he would suffer a loss since the reBAP is the lowest break-even limit. Thus, due to the objective of profit maximization, the BRP is willing to earn on being in the advantageous direction counteracting the system imbalance. *(Quadrant 2)*
Instance 3:	At events with the system long when there is a need for energy, and the BRP does not correlate with it, the trader will profit by notifying with the SMP range both exceeding the limits in case of wholesale and negative reBAP. The area between of these two levels is unfavorable for the BRP because the situation is profitable for BRP – the TSO pays to the BRP. So, there is an incentive to have higher prices for offsetting deviations, otherwise it will be breakeven or commercially unfeasible for the BRP to agree on the boundary prices, both wholesale and reBAP using the conventional markets. *(Quadrant 3)*
Instance 4:	A BRP with a coinciding state of his portfolio to the system, in other words aggravating it, will only have potential savings when the prices occur between the two boundaries. It means if SMP happened to be lower than wholesale and higher than negative reBAP upward, the BRP's surplus is better to have at a lower cost because this will be paid to the TSO as a loss. *(Quadrant 4)*

Once the upper and lower boundaries of the prices are known to the intelligent agents and based on the deviation quantities and submitted bids by flexibility providers, the best output for the control SP-BRP is computed. After the deals had taken place, the nomination is processed through the established platform with market rules MaBiS. Since the deviations are minimized, given the prices used were selected correctly, the BRP will pay lessened imbalance costs with such a trading mechanism at hand.

<u>**Implementation**</u>

The proposed market mechanism to enable efficient coordination of market players is thought to be realized via smart contracts written in a Turing complete Contract Oriented scripting language *Solidity* [133]. The interaction of smart contracts with *API* (Application Programmer Interface) via *JavaScript, or python, or go, or rust* allows to create diverse decentralized applications that can be integrated in the daily operations of BRP for portfolio bottom-up communication and timely optimization.

Specifically, in terms of the market calculation process (Figure 5-9), the SP's iMSys node first gathers deviations data, then locally performs all computing to avoid high transaction costs and coding complexity [134]. The manipulation of the sensitive data by the SP can be excluded with the next step of simplified lightweight checking by a smart contract on the blockchain. However, if the bottom peers of the group trust in their SP's market calculation, then the step of verification can be omitted. Once the correctness of the market calculation was verified, the payments are transferred by a payment smart contract (e.g. *Quotacoin*). The smart contract notifies all other iMSys that checking got approved.

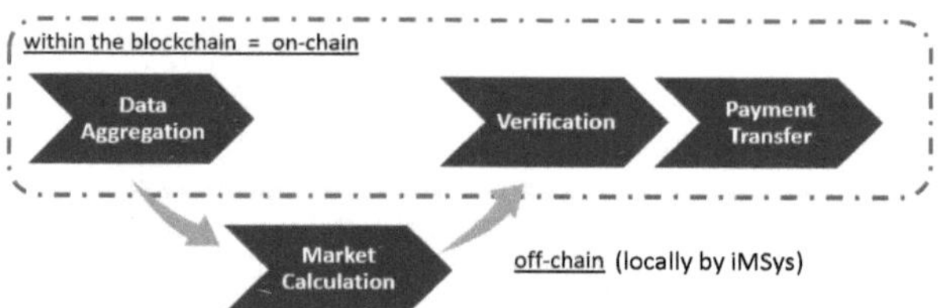

Figure 5-9: *Flow of the market processes in relation to the permission blockchain by BBG*

In the scenario of inter-chain communication, the market calculation must be done by a smart contract inside the public blockchain with their own tokens, while the permissioned blockchain operators (SPs) can manage their deviation-related data processes off-chain. The SPs offloads its deviations by inter-exchanging their native tokens with the uniform tokens inherent to the public blockchain.

This section presents where the blockchain technology can find its implementation in already existing decentralized data exchange platform for imbalance schedules communication in Germany.

According to [60], [135] the established data exchange supporting edi@energy data format with the basis of the UN/EDIFACT standard serves the main purpose of performing settlement process of energy imbalance between TSO and BRP. Worldwide, this data format also found its use in transmission of electronic data in a range of business communications. The most common documents exchanged through EDI are purchase orders, invoices and advanced shipping notices. All data flows are electronically captured and processed in highly automated and efficient manner. A more detailed description of process capabilities was already given in the fundamentals part of this thesis (p. 32).

The capacity of standardized and regulated German EDIFACT framework serves a great use for a range of involved parties such as network operators (TSOs and DSOs), suppliers, metering point operators and BRPs [65]. Although this flatfile format is 30 years old, the recent statistics of using EDIFACT framework show good results. For example, *Westnetz* (an RWE DSO) and *NetzeBW* (an EnBW DSO) sent over 500 million EDIFACT messages yearly that accounts and serves to approx. 12 million inhabitants (2015). Most German market partners accounting to 3000 regulated and unregulated companies are eager users of such leading syntax because its processes, messages and formats have always been fine-tuned due to fast developments in laws, acts and market policies.

However, the end-consumers are not having an access to their own metering data nor able to grant the access to data input themselves, nor to third parties who can help them make better informed choices [66]. On the other hand, all customer data is protected by strict German privacy law [136], [137], which also contradicts with blockchain features of data immutability (resistant to retractive modification of data). In line with the *EU-DSGVO* [138] (*EU General Data Protection Regulation* and the [139], it is clearly stated that according to *Articles 16, 17* and *20* individual's right to rectification, right to erasure (*"right to be forgotten"*), and also right to data portability, a person has right to demand for correcting, deleting or relocating the personal data he or she does not want to be publicly available in an unfavorable way. Unluckily, it directly contradicts the nature of blockchain irreversibility characteristics.

Still, this prohibitive design of the German 'decentralized' platform cuts off the active non-discriminatory involvement of the end-consumers, putting constrains against assisting in understanding their own electricity needs with the ability to reduce their consumption at critical times. Moreover, supplier ban on quotes to get known the limits and not bear penalties ex-post is how it is pictured today. Otherwise, when consumer-centric approach is applied [113] this could enhance transparency and add up to control over data sensitivity.

This means that since today imbalance data exchange is already automated enough it becomes rather easier and manageable to couple it with top-layer blockchain applications, instead of completely substituting. The *enterprise resource planning* (*ERP*) system interacts with blockchain layer via Oracle and SAP [109]. So, it is software-conforming to use complementary the blockchain platforms to augment and speed up settlement-related processes creating a synthesized record of information flows visible throughout the clearing phases. One might argue though, semantics in blockchain are not legally standardized yet which is the main reason why the EDI system cannot be coupled with promising blockchain technology [140].

All parties in the loop involved with sending transactions can receive alerts in real-time when requirements are met, or some abnormal profile fluctuations occur within the specified limits. This in turn will facilitate compliance and expedition of transaction settlement [141]. Checking for important data such as price differences, valid information and product availability dates could help in data validation and error detection, securing the parties beforehand and allowing for corrections to prevent costly delays further downstream. Moreover, EDI system's fundamental problem is in security – once a trading document is sent to a trading partner, there is no control over this data any longer, nor there is a way to track where it travels next down the chain [141]. Other methods of communication of schedules and planning-related documents between BRP and its balancing group customers include: *email, ISDN (Integrated Services Digital Network)*, and *FTP (File Transfer Protocol)*. The Table 5-7 below summarizes key blockchain pros against EDI's cons:

Table 5-7: *Slow-down factors of EDI transactions opposed to blockchain integration*

EDI	Blockchain
One-way and point-to-point messages	Many-to-many interactions
Popular networks (ANX, ENX, JNX)	Any new transactions are validated in real-time by smart contract rules with bidirectional flows
Risk of losing transactions	Shared ledger – no risk of losing transactions
Data tampering (hiding a data piece)	Impossible to hack and manipulate the data
Unstructured data is presented very differently by type of visibility and by tier of relationship	Shared information flow with transparency, ownership, auditability, and visibility
Storage of confidential data	No need of emails, spreadsheets, data feeds – all may be made available online
Usage of private keys	Eliminates the need for functional acknowledgements to confirm a delivery of a particular transaction

Erroneous data input by users	Real-time alerts
Disputes and repudiation (cancelling)	Preserves the transactions sequence in B2B process
	Increased energy product lineage

Not only the choice of technology format disadvantages the BGM, but also by using MaBiS platform the users are bound to wait overly extended time periods to execute all steps of data clearing and settlement as discussed in the *Chapter 2*, section 2.4.5. This delay from receiving requested data by TSO from DSO and BRPs takes as a general rule about 8 working weeks [38]. With the unique blockchain features of instant settlement [25] this process can be significantly shortened and hence optimized. The Figures 5-10 and 5-11 depict how lengthy and tedious the actual settlement chain is.

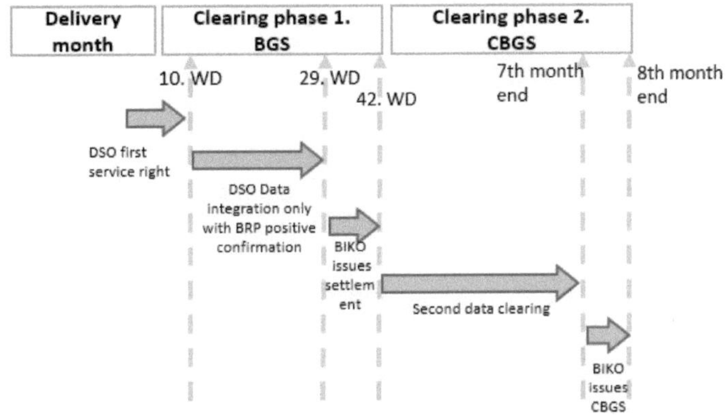

Figure 5-10: *The duration of clearing- and settlement deadlines according to MaBiS* [142]

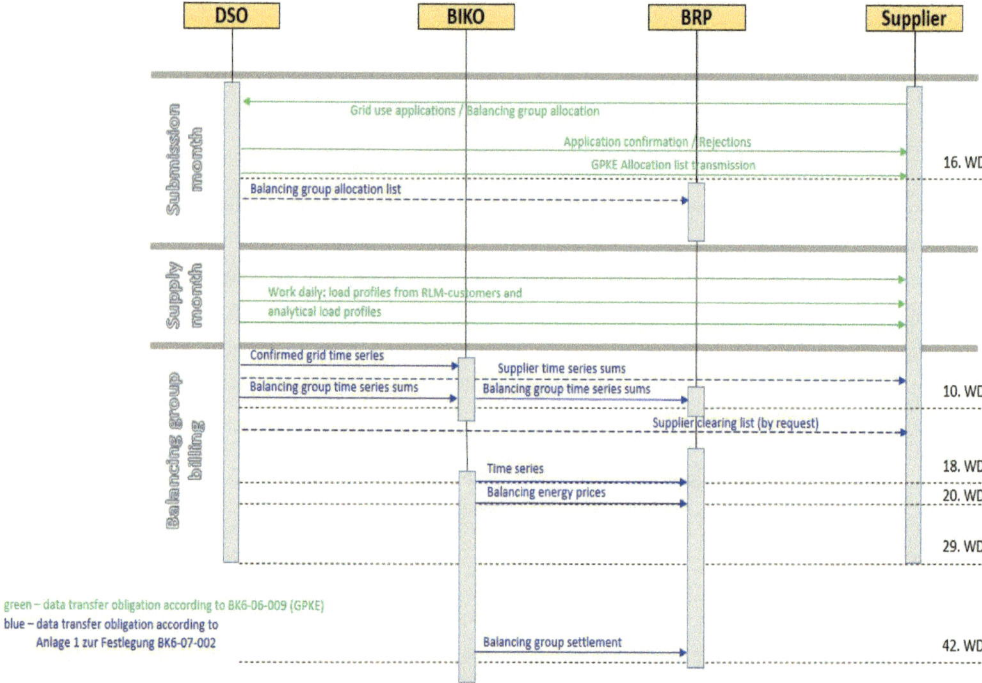

Figure 5-11: *Sequence diagram on the deadlines for balancing groups billing* [70], [143]

The second clearing phase is even assumed to be redundant and may be omitted soon according to [142] and [38]. Conversely, the Balancing Cooperation (in German - *BKK*) [144] points out that post-correction phase does not weaken the quality of balancing commitment, in opposite it helps TSO differ between balancing group settlements that took place whether before or after the delivery and only the allocation to BGs of energy quantities is changed and no new physical electricity flow thus can be released whatsoever. Plus, the option of post-delivery schedule correction reduces the billing balancing energy volumes, which in turn relieves TSO's liquidity management and minimizes the default risk for its business partners. Apart from this, other premises for necessary measures in the MaBiS processes are intensively discussed by grid operators and BKK focusing on the first clearing phase [142] (Table 5-8):

Table 5-8: *Current non-blockchain measures and its premises to improve the efficiency of MaBiS*

Desirable means	Potential effect
The withdrawal the second clearing phase only for the middle-term	It is believed to accelerate the actual settlement, but no proof and no testing was provided
Incentivizing the platform users for the quality of data provided during the first phase	This will in turn reduce risks of: 1. *wrong allocation* of BGs to metering points by DSO 2. *wrong single load profiles* made by DSO 3. *inaccurate aggregation* of BG total time series 4. *errors in data forwarding* toward TSO 5. *errors by aggregating* of BG total time series for the BG settlement purpose
Penalizing the delays or failure to meet the submission dates	This will foster the *punctual behavior* and potentially reduce the duration of overall clearing process
Preparing in anticipation of the 'smart meter' roll out	The premise of having by every energy asset an *'intelligent measuring system'* (iMS) will likely cut out participation of DSO in the processes (*discussed in the section 2.4.3*)
Re-assignment of the responsibility and tasks for performing balancing with the metered total time series	TSO will be responsible for remote reading, monitoring and acquiring of energy time series via iMS installed at each BG, followed by reconciliation check with submitted BRPs' schedules. This change will exclude the human factor and lots of intermediary steps by DSO and thus, promote a rather higher quality of the data and its timely acquisition

Furthermore, blockchain data formats allow to document, consolidate and account the energy data in a high quality and resolution providing not only the information on the moment of generation but also the lifecycle of its further utilization can be logged on the blockchain. The tracking of the energy attributes on the side will significantly improve the product visibility and ease access to the parties involved.

It is worthwhile to mention that 95% of data files of more than 8.000 BGs in total in Germany coming to TSO is foreign [38], meaning that TSO does not know how legitimate, authentic or fraudulent it can be. Generally, TSO is not capable of estimating the level of data quality and forwards the received data further through his consensus mechanisms that verifies and proves it eligibility after reviewing by several parties according to their internal policies. Already this indicates that a blockchain-based layer is a best fit for this process step. Because of the enormous number of BGs and DSOs participating, the involved parties for the data transmission cannot not be trusted. Moreover, the members of this organized data exchange process uniform data format. A *consortium* blockchain should be most suitable for the existing constellation of the roles in MaBiS framework. With TSO in the head since he has control over all data administration processes and stands in charge of the establishment of settlement toward BRPs. The Figure 5-12 identifies data exchange processes that involve several market parties, observes their trustedness and possible time delays in daily operations. Also, the choice between different types of blockchain network is checked.

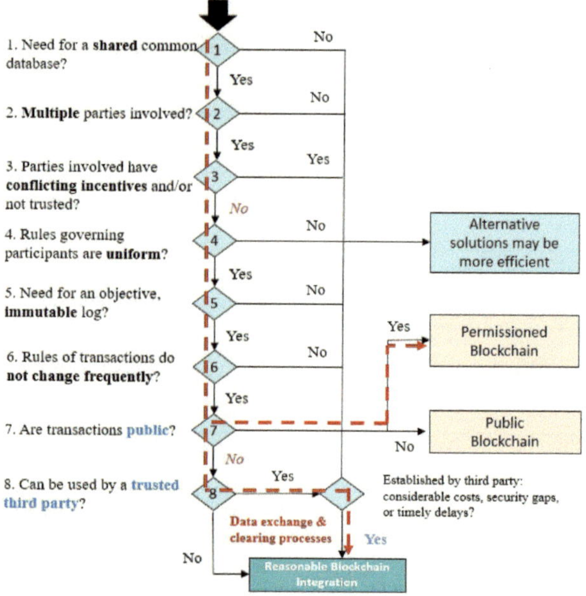

Figure 5-12: *Feasibility of blockchain incorporation in the balancing group management* [67], [87]

A similar blockchain feasibility analysis can be done for each individual BBGs as in the *section 5.1.2*.

5.2.1   Key technological benefits of blockchain in the MaBiS settlement processes

Once TSO has received the time series from DSO in the form of latest measurements from energy assets, TSO uses these energy values to audit and reconcile with the total time series that he got from BRPs. According to this TSO collects all BG data from different market partners and bases his final accounting.

The data flow and clearing phases could be automatically executed over the observable smart contracts, which execute simple encoded *'if-then'* functions in a secure and accountable manner. In return, this could optimize single processes of data exchange such that the settlement is accelerated [87] and [145]. The Table 5-9 highlights the main technological benefits [146] the incorporation of the blockchain technology in daily activities of BRPs and TSO may realize:

Table 5-9: *Key technological strengths of blockchain in applying with data exchange MaBiS*

I	The **real-time control** of the energy system [147] is deemed as the primary potential of blockchain. It allows administrating and documenting of feed-ins and draw-offs in real-time, a more precise grid operation and hence, a *lowered activation of balancing energy* is achieved by simplified clearing stages [27]
II	The **minimization of forecast errors** and **reduction in process steps**, meaning a lower amount of transactions to send, will decrease costs and simplify settlement process between TSO and BRPs on top of the MaBiS [147]. Additionally, it also may provide an interaction platform for the customers managed by BRPs. In turn, this would allow smaller in energy value transactions to be traded between parties due to reduced transaction costs which is not viable by far [148]. This opens up the new avenues of P2P trading for small energy communities locked in in the neighborhood or microgrids with limited competences in the traditional energy markets
III	The **lessened costs** come from the system size and transparency features brought by blockchain technology. Firstly, because central systems have more intermediaries involved due to various hierarchical structures, the management costs increase. Blockchain platforms are not bulky in size as many large organizations with a central server (e.g. *Amazon, Facebook* or *Google*) but rather set up by relatively small developer teams (5-20 persons). On the other hand, the low number of developers also may imply a common vulnerability issue when their codes are not often and properly examined by others [109]. Secondly, almost all blockchain projects are open-source (available at github.com), so they are publicly accessible with no fixed costs for purchasing, implementing and licensing software, nor further costly periodic maintenance is required. As a result, the transaction costs is diminished significantly and can be even none, while the transactions speed rises up. For example, up to *80% savings* in operational costs was achieved through the use of blockchain for billing within a virtual electricity community by *sonnen eServices* [109]. Only variable costs, such as electricity bills for mining, purchase of CPUs, building infrastructure and a choice of a blockchain network can add up to the transaction costs which differs by a particular case [149]

IV	**Enhanced security** is achieved through following technically ingrained features of blockchain:
	
	    i. improved data and processes integrity
	    ii. insurance against network failures
	    iii. transparency
	    iv. short duration of transaction execution
	    v. secure voting and consensus mechanism

V	**Reduction in electricity bill** for consumer [149] is achieved due to reduced redispatch costs with probably lowered grid fees, cutting in the intermediary cost share and through the *fair and transparent allocation of costs* by causation principle

5.2.2 Blockchain utility in financial transactions over MaBiS contracts

Alongside with the scheduling management, the BRPs and TSOs must have bank accounts for billing purposes after each settlement occasion on a monthly accounting basis. For the first rendering of accounting of balancing energy the deadline is the 42nd workday after the delivery month at latest. The second phase for sending a corrected invoice by TSO to BRP is up to the end of 8th month after the delivery month, while the BRPs have to submit their corrections up to the 7th month. So, the payments are split into two periods: first for the deviations before GCT or the delivery month and after, by the end of correction phase. Next kind of transactions usually take place between contractual parties [49]:

- Transfer of the monetary amounts (net) in the legal currency (€, EURO) on the total revealed deviation amounts. The money sum either is invoiced or credited to BRP's bank account, in cases of excess or shortfall depending on the system imbalance.

- VAT (value added tax) amount is also charged to each transaction

- Reasonable collateral is paid by BRP to TSO in cases of:
    - default in of due payments in a significant amount
    - outstanding debts in a material amount
    - irresponsible insolvency proceedings over the BRP's assets
    - failed to prove its creditworthiness

- Default interest charges are incurred in cases of a late payment alongside the costs for repeatedly requesting for the payment

- Distinct allocation of the balancing energy costs to the main BGs by a TSO.

The elimination of the time-consuming paperwork and bureaucracy from constant reconciliation of held databases against each other [13]. A small error or typo in one balance sheet (e.g. invoice) can be duplicated to copies of the document. Smart contracts can be designed such that invoice payment will be executed automatically, not involving manual labor. The blockchain integrates all vital information in one digital document continuously streamlining the bookkeeping, while all involved network parties can access it. The data kept in financial statements is the subject of audits, blockchain will allow automatic verification by proving the integrity of files leading to reduced costs and saving in time.

Turn of events: First, either the end-customer first write their financial transactions directly into a joint blockchain register held by a BRP and then TSO accesses these entries and bills directly the end-customer for the deviations. Or secondly, the BRPs do not open access to accounting, nor grant with rights of writing onto blockchain to end-customers. This means BRPs are the only ones who are accountable for the imbalance deviations which is the current German balancing market policy design.

Further, the cash flow between TSO and BRPs is deferred due to the arranged payments chain inherent to the system design with many steps. Namely, there are many challenges that lag and appreciate the transaction. The implementation of the blockchain layer on top of the protocols used within the MaBiS may significantly simplify step processes and bypass the banking middle men. Coupling these two mediums will synergize in the instant settlement with improved visibility that shall increase cash liquidity of commercial parties and decrease debt rate. The key premises that were confirmed and discussed by many research institutions [27], [87], [109] are pinpointed in the Table 5-10:

Table 5-10: *Economical MaBiS downsides and key promises that blockchain can cultivate into it*

Drawbacks		Promises
involves many workarounds	❖	- ramping up the speed of exchange - exclusion of transaction tampering and invalid removals - decline in the transacting backlog
high overhead costs	❖	- credit risk minimization
error-prone costly back-office procedures of a transaction	❖	- process integrity and inter-operability by standardizing blockchain data formats
frictional costs with large effort by dealing with central banking systems	❖	- reduction in the overall costs - disregard of the minimum transacting capital prerequisite
constant reporting to the regulator required by utilities	❖	- reliable and easily available data - improvement in the bulky auditing due to close to real-time record verification - tokenization of energy values in trend with digitalization

Also, the potential of blockchain recently has gained traction and increasing recognition for faster switching between energy suppliers [150] defined currently in the switching processes *GPKE* (*"Geschäftsprozesse für die Kundenbelieferung mit Elektrizität"*), although the actual thesis studies blockchain potential only for MaBiS settlement processes. However, it is worth to mention the main contribution of the referenced research work because it relates to the BGM area to a similar extent as MaBiS does. The proposed blockchain-based approach by [150] allows to substantially lower the processing time to below than 15 minutes interval. The current GPKE switching processes can last up to 15 days due to the involvement of multiple inefficient intermediaries. Secondly, the mapping feature of Ethereum Blockchain platform utilized in the approach avoids other complex and partially redundant verification processes as well as manual checks and validating by data transmission leading to lengthy and extended procedures.

5.2.3 Potential blockchain Use Case in the sub-BG supervision

Sub-BGs are managed under a main BG. Only initial *'pooling'* BG is financially accountable by the BGM contract for the total net deviations. Main BG has its own responsible manager (BRP) for accounting who redistributes deviation costs to underlying sub-BGs based on internal agreement [59].

The sub-balancing groups can form cascades downstream creating multitude of sub-sub-…-balancing groups (Figure 5-13). Attaching BGs steered by foreign BRPs is discouraged to not overcomplicate the billing [151]. Initially before 2012, all sub-BGs down the chain at different cascade levels would communicate their schedules directly with TSO, whereas today due to a rapid growth in quantity they informally communicate their energy use, hence deviations, only to the upper *link-BG*.

The sub-BGs are not obliged to ensure quarter-hourly balance performance by the BGM contract as the main BGs are, and thus do not send the forecast schedules to the TSO. Sub-BGs re-allocate their balance deviations to the main upstream BG manager who tracks and takes measures to offset them and stands financially responsible for the deviation payments. The main BG breaks down all sub-layered BGs' energy values in the schedules such that the TSO knows exactly who caused one or the other imbalance.

Yet, every sub-BG manager also has its bank account similar to the main BG. After receiving a paycheck from TSO for final settlement by 8th month after the delivery, the accountant of the main BG sends invoice/credit notes to their sub-BG mangers too based on the costs-by-cause principle. In other words, the top BG manager bills according to their contribution to the system imbalance. The internal redistribution of the payments is not regulated by the BGM contract, yet there is only an agreement that states that sub-BGs are not against to re-allocate their deviation costs to the top-standing BRP. Only main-BG BRP economically estimates these costs which complexity amplifies by spreading with sub-class groups. The Table 5-11 reveals the truth about this long-chain-billing approach:

Table 5-11: *Merits and drawbacks of pooling sub-BGs in the cascades*

−	Those deals are quite intransparent and untraceable, as well as this 'internal relationship' has its costs. The sub-BG does not have an economical advantage in the balancing group settlement processing. This is because electrical balancing energy has symmetrical single pricing, meaning by purchasing and selling the price is the same, no differentiation
+	Internal ridding of deviations is possible, and it is therefore highly reasonable to provide management services to deal with settlement from main-billing BG A to the sub (-pooled) BG B. Sub-BGs can offer advantages in the main settlement or extra desirable in-house operational transparency. This is where blockchain technology can provide a platform for already decentral communication of energy data and billing (see Figure 5-13)

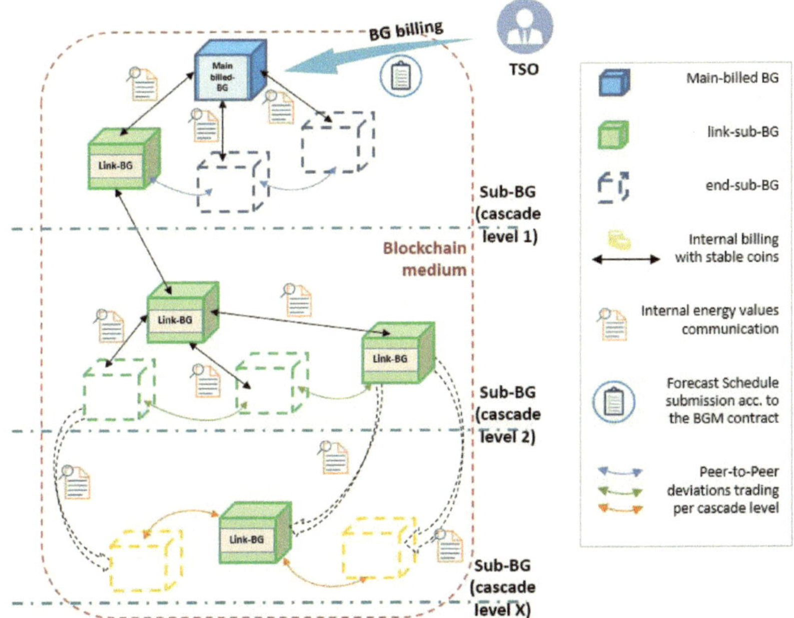

Figure 5-13: *A blockchain Use Case to resolve a hierarchical structure of sub-BG management*

With several cascade levels ruled by the main-billed BG, this structure presents a blockchain use case. So, it is mainly the job of the BRP leading the main BG in the cascade head to keep balancing sheets clean and have efficient optimization and administration of downstream deviations to avoid wrong and costly re-allocation of imbalance costs to its sub-BG managers. This whole management system by the BRP of his sub-BGs in virtual accounts highly resembles how actually blockchain is structured. It seems that current complex management with insecure communication means of BRP portfolio (discussed in *section 5.2*) can be improved by user-friendly and transparent blockchain-based applications with higher degree of independence.

6 Conclusions

6.1 Discussion & Conclusion

Summary

Based on an extensive economical evaluation of the balancing cost development, thorough investigation of the P2P-trading regulatory environment, this thesis proposed the simulation of most probable scenarios for the incorporation of the blockchain components in different areas of electricity balancing management in Germany. Mainly caused by the rise of intermittent renewables from energy transition, power outages and sudden changes in consumption, the detrimental fluctuations in the energy system and trading strategies on the energy markets threaten the security of energy supply. Especially, inaccurate load forecasting leading to higher financial losses, ever growing complexity of internal portfolio optimization lacking in transparency for end-customer, alongside with high volatility of imbalance prices (reBAP) makes the main energy market players – BRPs – seek for various innovative tools. The tools that appreciate regulator's main policy for electricity balancing to strengthen even more and uphold BRP balancing commitment with the neutral state of portfolio. However, the actual organization of the German imbalance settlement mechanism does not allow for the active participation of end-customer in helping stabilize the system due to the balancing market design, regulatory barriers and weak economic incentive. The German regulation promotes rather bulky grid updates encouraging flat consumption ruling out most benefits of demand response programs. Therefore, it was interesting to find optimally acceptable and regulatory-conform solutions of efficient offsetting of deviations incurred to financially accountable business BRPs. Evaluating economical incentive of BRP and testing various business models with technological extensions following the trend of energy digitalization in Germany led to the development of prosumer-integrated service model that achieves maximum utilization rate of already ample energy resources. This service model rather counteracts the prosumer unpredictable behavior and aims to revitalize silenced end-consumers. As a result, it enables the desired peer-to-peer timely counteraction to the system imbalance under soft BRP guidance with price signals at hand.

Chapter 2. Once the potential of the blockchain technology for energy balancing purposes and settlement got recognized, it was very important was to cover essential principles of regulation market design and then provide the fundamental knowledge of the balancing group management in Germany as the first step. This thesis discovered a clear difference in commercial responsibilities for the activation of control energy between BSP and BRP, followed by BRP classification by core principles of balance group structure, their inter-communication and legal obligations and rights stated by market rules 'MaBiS'. Especial attention was given to the established data exchange platform ESS over which the focus element of balance groups – *forecast schedule* – is processed and reconciled with realized plans. In this work it was found how inefficiently lengthy, tedious and mediator-braking the settlement process of caused deviations is. It was estimated whether the blockchain fits the purpose of cost-effective daily operations of BRP and strategically cultivate the added-value benefits of accelerating the settlement process instilling transparency in the origin of deviations.

Chapter 3. The next logical step was to unveil the country-specific economical parameters the market players are oriented around. Initially, the performed visualization of the country-wide deviations in consumption profiles revealed the quantitative mismatch of the realized values from prognosed with a clear pattern in seasonality factor as well as the dominancy in undersupply events in 2017 (9 GWh short vs. 3,9 GWh long). With the goal of estimating the historic trend of total balancing cost separating the regulated system from business BRP, the sufficiency of current incentive for the BRP to reduce own imbalance was examined.

In essence, a serious impact of the German BRP cost is nearly always caused by the absence of information on the sign of balancing energy volume in real-time with a delayed publication of imbalance prices. Specifically, the German costs allocation rule and pricing mechanism (symmetric) translates into *ex post* accounting settlement of associated imbalance costs to BRPs with deviated day-ahead commitments from only activation of balancing capacity. While the technical costs associated with its reservation are socialized via grid charges to end-consumers, thereby hiding the marginal cost component for the reBAP calculation that potentially would enhance incentive for BRP. Moreover, the imbalance price spread being a key incentive for BRPs to improve forecast quality showed twice higher values for the system long in 2017 meaning an increased tendency for under-contracting the system (short) by BRPs. Plus, it was spotted that a perverse incentive sent by the system has increased in events of system undersupply thereby misleading the BRPs in balancing. All of this implies higher occurrence of payments from BRPs to TSOs, but it is still not clear whether or not the internal operational profits, which is confidential information, outweighed these allocated imbalance costs by every individual party to claim a profitable operation.

As a second approach the publicly available data on the German BRP net costs were acquired and studied to get a more precise insight about how stricter became the incentive to uphold balancing commitment by BRPs with years. The cost comparison analysis showed a steadily gradual decline in system balancing net costs from activation of control reserves, while the re-allocation of costs after settlement with BRPs did not follow the system trend but since past two years rather started growing noticeably on the contrary.

Since the thesis seeks an alternative to the existing methods of offsetting the imbalance deviations, the blockchain type of market was considered. With existing centralized trading options at hand to offset deviations for BRPs it was interesting to see their decision making on one or the other type of market. After comparing it became clear that there is a very weak correlation between the day-ahead prices and highly volatile and unpredictable German imbalance prices leading to an inefficient arbitrage. As a result, the attempt for risk-averse BRP to speculate on more attractive prices by market type was rather excluded in the current balancing market design with unknown reBAPs in short-term. Therefore, following the risk management assessment it was decided that a blockchain-based market with a similar hedging idea of deviations' ridding with timely price signals can certainly compete with conventional solutions. Namely, due to smart contracts functionality, increased digital product lineage and higher adaptability of innovative P2P business models.

Chapter 4. However, for a more accurate adaptability of P2P trading using blockchain in Germany, the regulatory environment was checked in order to prove the feasibility of the concept. First, the aggregator services were considered as the most viable tools for offsetting deviations by pooling several types of energy assets and performing internal scheduling and optimization processes. Because of several regulator disincentives and conflict of interest between the supplier-BRP and an independent aggregator with DR services, it was decided to not link a potential blockchain business case to the regulation market. In particular, aggregators must first acquire a number of permissions for legal operation in Germany.

After the excessive research on the existing market solutions for balancing group operation with blockchain in this thesis it was determined that it is difficult for prosumer to market its energy P2P in a decentral fashion due to several legal barriers and bureaucratic hurdles. The profound stumbling block for the end-customer was to register itself as a Balancing Responsible Party, which implies for a private individual – a number of nearly impossible admissions to acquire with further strict compliance for grid and energy law.

Taken into account the power of utility companies in Germany who have already set all of these political, legal requirements and tons of experience in dealing with deviations having the energy markets access, in this thesis was decided to opt for a *service model*. Essentially, it enables a C2C-prosumer as system integrated for trading over the blockchain-based platform under legal covering of a service provider.

The scheme was elaborated on the classification of service provider resulted in five categories: in-house aggregator, independent aggregator, power utility company, BGM-as-a-Service, and E-community manager. Under this setting a service provider as a BRP forms a blockchain balancing group network providing its customers with technical equipment necessary to connect to their local flexibility market. With P2P market at disposal, a more efficient internal BG portfolio management is achieved, while external is also possible over the inter-chain communication.

Chapter 5. Once the way for enabling of prosumer that overcomes the regulatory issues was found, it was interesting to simulate the business interaction of market players within the service model. Therefore, several scenarios for P2P trading, such as *intra-Group, in-Group,* and *inter-Group* were developed based on the established concept of balancing groups in Germany. Firstly, the end-customer blockchain-based market was proposed where based on quota-mechanism energy consumers could market their deviations under price signals from the network service provider. Each participant is equipped with an intelligent measuring system which serves as a control smart device that maintains blockchain protocols and simultaneously is used for remote energy monitoring and dynamic pricing. The peers of the network can offer their capabilities of switching on/off load expressed in *quotas* on the secondary market. The quota units are proposed as differentiated product that is assembled of multiple properties defined by each blockchain node. The proposed market architecture was decided in favor of proof of authority consensus mechanism executed on the custom permissioned network. Due to alarm management functions time-prioritizing and filtering by the deviation difficulty, service providers receive nearly real-time signals. Such flags inform on the state changes in balance of their portfolio. Based on this a logical pricing algorithm is run to determine the most economically efficient price range for the BRP to resolve the occurred deviation and bring the portfolio in the neutral state.

After internal portfolio optimization was performed, the BRPs were granted with further stage of inter-operation over the public blockchain. This was elaborated by bridging inter-chain solutions enabling exchange of remained deviations on the level of different blockchain BG networks. The proposal is rounded off with a possible steps flow of the secondary market showing different variants for data aggregation and market calculation.

In conclusion, the last section focused on the feasibility of incorporating the blockchain components in the established data exchange in the chain BRP-TSO-DSO operating under the market rules 'MaBiS'. Through the existing platform the transmission of forecast schedules is realized. In the spotlight the applied data format EDIFACT was opposed to the blockchain benefits to find out how it could be beaten by a more modern technology. Real-time data stream combined with tamper-proof resistance and automated process execution with enhanced transparency proved the settlement process optimization a reasonable use case for blockchain. The vehicle of BGM resulting in commercial settlement allocates each mismatch to a respective BG with subsequent billing based on causation principle. This process calls for the coupling of the blockchain technology with the premise of improving technical data quality, exchange integrity, security and transparency. In turn, automated process steps can be achieved leading to higher efficiency in the BGM, shortening the settlement process steps and hence, reduced demand for calling control reserves could be attained.

6.2 Recommendations and outlook

Admittedly, the blockchain technology is relatively new but already has instilled very promising innovations in the energy sector, especially in the area of electricity balancing. Although this thesis was investigating only balancing group management aspects predominantly related to economical cost and risk of deviations, the congestion management problem calls for rather more urgency – a 14-fold increase of costs over the last 6 years [6]. The blockchain-featured platforms can provide the ability of efficient demand side management that is a subject of DSO-TSO communication with a grid optimization problem. Proper regulatory changes in the independent aggregator businesses can instigate timely marketing of a variety of distributed energy assets directly in regulated market. Only standardized intelligent metering system could facilitate this that is a mainstream policy of regulator for the next year.

Similarly to the balance group management area, a host of blockchain use cases can be investigated on the regulation market side. Firstly, in pre-qualification processes for the providers of control capacities a decentralized secondary market for offering a place in the merit order list can be built using blockchain. It can bring higher transparency that is a key concern in this area of today. Secondly, a more efficient control power optimization can be achieved through the excluding misactivation of assets with smart contracts. Lastly, a cross-border deviations exchange over a decentralized platform with blockchain extensions can promote a more decentral collaboration with higher economical outcome.

For strengthening BRP's incentive to be actively engaged in restoring system balance even further ahead, the value of avoided control energy activation has to become ubiquitously available for the market players. This reference point would allow a more accurate economical estimation of a trade-off that BRP faces every day. A change in regulation is then inevitable, once the publication of system's balancing volume sign becomes accessible online in nearly real-time, a more accurate estimation for BRP portfolio can come true. The inherently tedious allocation of costs responsibilities protracts the regulator in changing the rules. When these calculations of utilized balancing energy are transparent nearly real-time with the help of blockchain, then market players could react sharper to adjusting notified schedules and true market value of electricity shall reveal.

In its nature shared economy concept is driven to exchange goods and services based on preferences and capabilities. End-customers having a potential to provide timely activation of their energy assets should not be silenced nor neglected. When the accurate matching in line with price signals organized, a new stream of revenues could attract a range of energy assets into the balancing processes. Also, by introducing different bonus programs appreciating the readiness to 'turn on', as well as network fee discounts could and shall enable the system integration of the end-customer or deviations-aware energy communities. Still, the proposed interaction of such assets over blockchain must be technically implemented and thoroughly tested by a real service provider like a utility company or a business BRP.

For further improvement of the customer-level trading using blockchain, an efficient measure could be rethinking grid tariffs and enabling dynamic prices that shall offset the current economical infeasibility of executing P2P prosumer trades. Another technique of mapping separate blockchain balancing groups is definitely a way to increase the chance of finding a counterparty to cancel out the deviations at a shorter notice. A direct exchange of discrepancies in schedules by parties on various levels leads to fulfilling the regulator's main objective of securing the equilibrium between supply and demand. The interconnectedness provided by blockchain tech paves the way and predestined to attain this goal.

An additional blockchain use case was discovered in the management of sub-BGs field. Because the internal interaction is not regulated by the BGM contract but governed under the informal agreements by the billing BG, the complexity is growing with the number of cascades falling downstream. A better transparency and eased management in this area can be provided by the blockchain tech that serves as a bookkeeping tool reflecting all recorded positioned stored and fixed in the ledger. Instead of one blockchain layer for the whole portfolio use, their nesting at each cascade can preserve confidentiality. Further, these entries could be of great use for a proposed secondary market mechanism.

In the future work it will be interesting to run pilot tests with processes similar currently labor-intensive established data exchange with market rules MaBiS. The studied processes can be combined with smart contracts that will significantly offload reconciliation jobs carried by TSO, BRPs and DSOs. But this can only be then achieved when there are regulatory-approved standards for the syntax and semantics utilized within the blockchain space that could assure a seamless and more efficient operation compared to what exists today.

Bibliography

[1]   ENTSO-E Transparency Platform. "Financial Expenses and Income for Balancing [17.1.I]."
        (2018) . Available:
        https://transparency.entsoe.eu/balancing/r2/financialExpensesAndIncome/show?name=&defaul
        tValue=false&viewType=GRAPH&areaType=MBA&atch=false&dateTime.dateTime=01.01.2
        015+00:00%7CUTC%7CMONTH&dateTime.endDateTime=01.11.2018+00:00%7CUTC%7C
        MONTH&marketArea.val. [Accessed: 25-Nov-2018].

[2]   BMWi. "An electricity market for Germany' s energy transition." In: *White Paper by the
        Federal Ministry for Economic Affairs and Energy* (2015) .

[3]   Ecofys and Frauenhofer-IWES. "Smart-Market-Design in deutschen Verteilnetzen in
        deutschen." In: *Agora Energiewende* (2017) . Available: https://www.agora-
        energiewende.de/fileadmin2/Projekte/2016/Smart_Markets/Agora_Smart-Market-
        Design_WEB.pdf. [Accessed: 04-Jan-2019].

[4]   M. Klobasa *et al.* "Monitoring der Direktvermarktung von Strom aus Erneuer-baren Energien
        Vorbereitung und Begleitung bei der Erstellung eines Erfahrungs berichts gemäß 97
        Erneuerbare- Energien--Gesetz." In: *Fraunhofer ISI, Fraunhofer IEE, IKEM, THI* (2017) .
        Available: https://www.erneuerbare-
        energien.de/EE/Redaktion/DE/Downloads/Berichte/monitoring-direktvermarktung-strom-ee-
        quartalsbericht-12-2017.pdf?__blob=publicationFile&v=2. [Accessed: 16-Nov-2018].

[5]   L. Hirth and I. Ziegenhagen. "Balancing Power and Variable Renewables: Three Links." In:
        Renewable & Sustainable Energy Reviews (2015) . Available:
        http://www.sciencedirect.com/science/article/pii/S1364032115004530. [Accessed: 23-Dec-
        2018].

[6]   M. Joos and I. Staffell. "Short-term integration costs of variable renewable energy: Wind
        curtailment and balancing in Britain and Germany." In: *Renewable and Sustainable Energy
        Reviews* (2018) .

[7]   P. Olivella-Rosell *et al.* "Local flexibility market design for aggregators providing multiple
        flexibility services at distribution network level." In: *Energies* (2018) .

[8]   B. Neupane, T. B. Pedersen, and B. Thiesson. "Utilizing Device-level Demand Forecasting for
        Flexibility Markets - Full Version." (2018) . Available: http://arxiv.org/abs/1805.00702.
        [Accessed: 28-Sep-2018].

[9]   J. Huber, N. Klempp, and K. Hufendiek. "An interactive Online-Platform for Demand Side
        Management." In: *Proceedings of the Ninth International Conference on Future Energy
        Systems* (2018) .

[10]   O. Langniß, F. Förster, and T. Brenner. "OLI Systems GmbH." (2018) . Available:
        https://www.my-oli.com/de/. [Accessed: 05-Jan-2018].

[11]   P. Kraus. "Master thesis - Architectural Analysis of Blockchain-based Approaches to a
        Decentralized Energy Management System." University of Stuttgart (2018).

[12]   M. Yahya. "Master thesis - Decentralized Green Energy Based Certification on the top of a
        Web3 Based Energy Geo Browser." Hochschule für Technik Stuttgart (2018).

[13]   M. Merz. "Potential of the Blockchain Technology in Energy Trading." In: *Blockchain technology Introduction for business and IT managers* (2016) .

[14]   E. Mengelkamp *et al.* "A blockchain-based smart grid: towards sustainable local energy markets." In: *Computer Science - Research and Development* (2018) .

[15]   E. Mengelkamp *et al.* "Designing microgrid energy markets: A case study: The Brooklyn Microgrid." In: *Applied Energy* (2018) . Available: https://doi.org/10.1016/j.apenergy.2017.06.054. [Accessed: 11-Jul-2018].

[16]   J.-C. Juncker. "COMMISSION REGULATION (EU) 2017/2195 of 23 November 2017 establishing a guideline on electricity balancing." In: *Official Journal of the European Union* (2017) . Available: https://eur-lex.europa.eu/legal-content/EN/TXT/?uri=uriserv:OJ.L_.2017.312.01.0006.01.ENG&toc=OJ:L:2017:312:TOC#d1 e259-6-1. [Accessed: 21-Dec-2018].

[17]   H. Bontius *et al.* "USEF: WORKSTREAM An introduction to EU market-based congestion management models." (2018) . Available: https://www.usef.energy/app/uploads/2018/04/USEF-DSO-WG-report-Overview-Market-based-congestion-management-v1.00-FINAL.pdf. [Accessed: 21-Aug-2018].

[18]   Schleswig-Holstein Netz AG and ARGE Netz GmbH & Co. KG. "ENKO – Das Konzept zur verbesserten Integration von Grünstrom ins Netz." (2018) .

[19]   A. von Hammer-stein, Christian von Bremen and P. Roegele. "AGGREGATOREN - FESTLEGUNG BESCHLOSSEN." . Available: https://raue.com/aktuell/branchen/energie-rohstoffe-und-klimaschutz/aggregatoren-festlegung-beschlossen/. [Accessed: 20-Dec-2018].

[20]   P. De Filippi. "What Blockchain Means for the Sharing Economy." In: *Harvard Business Review* (2017) . Available: https://hbr.org/2017/03/what-blockchain-means-for-the-sharing-economy. [Accessed: 24-Jul-2018].

[21]   S. Nakamoto. "Bitcoin: A Peer-to-Peer Electronic Cash System." (2008) . Available: https://bitcoin.org/bitcoin.pdf. [Accessed: 20-Jun-2018].

[22]   M. N. Luke *et al.* "Blockchain in Electricity: a Critical Review of Progress to Date." (2018) .

[23]   VDI Technologiezentrum. "Blockchain - Eine Technologie mit disruptivem Charakter." (2018) . Available: https://www.vditz.de/fileadmin/media/news/documents/Blockchain_-_Eine_Technologie_mit_disruptivem_Charakter.pdf. [Accessed: 29-Oct-2018].

[24]   Wikipedia. "Blockchain." . Available: https://en.wikipedia.org/wiki/Blockchain. [Accessed: 26-Oct-2018].

[25]   M. Iansiti and K. R. Lakhani. "The Truth About Blockchain." In: *Harvard Business Review* (2017) . Available: https://hbr.org/2017/01/the-truth-about-blockchain#comment-section. [Accessed: 13-Dec-2018].

[26]   J. Bertsch *et al.* "Disruptive Potential in the German Electricity System – an Economic Perspective on Blockchain." In: *ewi Energy Research & Scenarios GmbH* (2017) . Available: http://www.ewi.research-scenarios.de/cms/wp-content/uploads/2017/07/Disruptive_Potential_in_the_German_Electricity_System_–_an_Economic_Perspective_on_Blockchain.pdf. [Accessed: 01-Sep-2018].

[27]   PricewaterhouseCoopers. "Blockchain – an opportunity for energy producers and consumers?" In: *PwC global power & utilities* (2016) . Available: www.pwc.com/utilities.

[28]   H. Natarajan, S. Krause, and H. Gradstein. "Distributed Ledger Technology (DLT) and Blockchain: FinTech Note 1." (2017) . Available: http://documents.worldbank.org/curated/en/177911513714062215/pdf/122140-WP-PUBLIC-Distributed-Ledger-Technology-and-Blockchain-Fintech-Notes.pdf. [Accessed: 17-Dec-2018].

[29]   D. Tapscott and A. Tapscott. "Blockchain Revolution: How the Technology Behind Bitcoin Is Changing Money, Business, and the World." In: *Penguin Publishing Group* (2016) . Available: https://www.youtube.com/watch?v=Pl8OlkkwRpc. [Accessed: 20-Dec-2018].

[30]   M. Swan. "Blockchain: Blueprint for a New Economy." In: *O'Reilly Media* (2015) .

[31]   K. Panetta. "5 Trends Emerge in the Gartner Hype Cycle for Emerging Technologies, 2018." (2018) . Available: https://www.gartner.com/smarterwithgartner/5-trends-emerge-in-gartner-hype-cycle-for-emerging-technologies-2018/. [Accessed: 23-Dec-2018].

[32]   D. Yaga *et al.* "Draft Blockchain Technology Overview (NISTIR-8202)." (2018) . Available: https://csrc.nist.gov/publications%0Ahttps://csrc.nist.gov/CSRC/media/Publications/nistir/8202/draft/documents/nistir8202-draft.pdf. [Accessed: 20-Dec-2018].

[33]   V. Buterin. "A NEXT GENERATION SMART CONTRACT & DECENTRALIZED APPLICATION PLATFORM." (2013) . Available: https://cryptorating.eu/whitepapers/Ethereum/Ethereum_white_paper.pdf. [Accessed: 02-Jan-2019].

[34]   S. Albrecht *et al.* "Dynamics of Blockchain Implementation - A Case Study from the Energy Sector." In: *Proceedings of the 51st Hawaii International Conference on System Sciences* (2018) . Available: https://scholarspace.manoa.hawaii.edu/handle/10125/50334. [Accessed: 06-Aug-2018].

[35]   The Linux Foundation. "Hyperledger Architecture, Volume 1." (2017) . Available: https://www.hyperledger.org/wp-content/uploads/2017/08/Hyperledger_Arch_WG_Paper_1_Consensus.pdf. [Accessed: 03-Jan-2019].

[36]   S. Hartnett *et al.* "The Energy Web Chain." (2018) . Available: http://www.energyweb.org/papers/the-energy-web-chain. [Accessed: 03-Jan-2019].

[37]   M. N. Luke *et al.* "Blockchain in Electricity: a Critical Review of Progress to Date." In: *NERA Economic Consulting, Eurelectric* (2018) . Available: http://www.energie-nachrichten.info/file/01 Energie-Nachrichten News/2018-05/80503_Eurelectric_1_blockchain_eurelectric-h-DE808259.pdf. [Accessed: 18-Dec-2018].

[38]   S. Bauernschmitt. "Marktprozesse für die Bilanzkreisabrechnung Strom (MaBiS)." In: *BDEW-Informationstag „Einführung in die Marktprozesse des deutschen Energiemarktes" by TenneT TSO GmbH* (2015) . Available: https://www.tennet.eu/fileadmin/user_upload/The_Electricity_Market/Rules_and_procedures/German/mabis/Marktprozesse_für_die_Bilanzkreisabrechnung_Strom__MaBiS__Nürnberg_Dezember_2015.pdf. [Accessed: 09-Jun-2018].

[39]   Wikipedia. "Synchronous grid of Continental Europe." (2018) . Available:

https://en.wikipedia.org/wiki/Synchronous_grid_of_Continental_Europe. [Accessed: 21-Dec-2018].

[40] Tractebel Engineering. "Study of the interactions and dependencies of balancing markets, intraday trade and automatically activated reserves." (2009) .

[41] ELIA. "STUDY ON PAID-AS-CLEARED SETTLEMENT FOR AFRR & MFRR ACTIVATED ENERGY." (2017) . Available: http://www.elia.be/~/media/files/Elia/publications-2/Public-Consultation/2017/20171023_Paid-as-cleared-FRR-study-v4.pdf. [Accessed: 10-Oct-2018].

[42] L. Giles. "Monitoring Report 2017." In: *Bundesnetzagentur für Elektrizität, Gas, Telekommunikation, Post und Eisenbahnen; Bundeskartellamt* (2017) . Available: https://www.bundesnetzagentur.de/SharedDocs/Downloads/EN/Areas/ElectricityGas/Collectio nCompanySpecificData/Monitoring/MonitoringReport2017.pdf?__blob=publicationFile&v=2. [Accessed: 14-Nov-2018].

[43] REWAG - Regensburger Energie- und Wasserversorgung. "Regelenergie: Vermarkten Sie Ihre Anlage mit der REWAG." . Available: https://www.rewag.de/geschaeftskunden/grosskunden/stromerzeuger/regelenergievermarktung. html. [Accessed: 03-Jan-2019].

[44] NextKraftwerke. "Was ist Regelenergie?" (2018) . Available: https://www.next-kraftwerke.de/wissen/regelenergie. [Accessed: 02-Oct-2018].

[45] J. Greunsven *et al.* "Market review 2017, Electricity market insights." In: *TenneT, IAEW* (2017) . Available: https://www.ensoc.nl/files/20180405-market-review-2017-bron-tennet.pdf. [Accessed: 08-Sep-2018].

[46] NextKraftwerke. "Und sie bewegt sich doch." (2015) . Available: https://www.next-kraftwerke.de/energie-blog/elastische-stromnachfrage. [Accessed: 14-Dec-2018].

[47] Wikipedia and Tennet TSO. "Balancing Groups." (2018) . Available: https://de.wikipedia.org/wiki/Bilanzkreis. [Accessed: 18-Dec-2018].

[48] ENTSO-E. "Energy Identification Code (EIC)." (2018) . Available: https://www.emissions-euets.com/internal-electricity-market-glossary/1712-eic-code. [Accessed: 27-Dec-2018].

[49] BNetzA. "Balancing Group Contract on the Management of Balancing Groups." In: *TenneT TSO GmbH* (2016) . Available: https://www.tennet.eu/fileadmin/user_upload/The_Electricity_Market/German_Market/Grid_c ustomers/contracts/BNetzA-BKC_englisch.pdf. [Accessed: 20-Dec-2018].

[50] TenneT TSO GmbH. "Bilanzkreisvertrag über die Führung von Bilanzkreisen." (2016) . Available: https://www.tennet.eu/fileadmin/user_upload/The_Electricity_Market/German_Market/Grid_c ustomers/contracts/Bilanzkreisvertrag_TenneT_2016.pdf. [Accessed: 21-Dec-2018].

[51] K. Goldammer. "Einführung in das Bilanzkreismanagement." In: *Reiner Lemoine Institut* (2016) . Available: https://www.strommarkttreffen.org/1.1-2016-09-02-Goldammer-Bilanzkreismanagement.pdf. [Accessed: 04-Jan-2019].

[52] Stadtwerke Unna GmbH. "VDEW-Lastprofile." (2014) . Available: http://www.gipsprojekt.de/featureGips/Gips;jsessionid=FB97EF608D23211289E999CA55C80

8B9?SessionMandant=sw_unna&Anwendung=EnWGKnotenAnzeigen&PrimaryId=133029& Mandantkuerzel=sw_unna&Navigation=J. [Accessed: 20-Jun-2018].

[53] de.wikipedia.org. "Standardlastprofil." (2018) . Available: https://de.wikipedia.org/wiki/Standardlastprofil. [Accessed: 17-Aug-2018].

[54] C. Möller, S. T. Rachev, and F. J. Fabozzi. "Balancing energy strategies in electricity portfolio management." University of Karlsruhe and KIT (2011).

[55] A. Witthohn. "Konzept für die Reduzierung der Risiken der ÜNB durch betrügerische Fahrplananmeldungen – Fahrplanabwicklungskonzept –." In: *Bilanzkreiskooperation* (2018) .

[56] M. Scherer, O. Haubensak, and T. Staake. "Assessing distorted trading incentives of balance responsible parties based on the example of the Swiss power system." In: *Energy Policy* (2015) . Available: https://www.uni-bamberg.de/fileadmin/eesys/PDF/Assessing_disorted_trading_incentives.pdf. [Accessed: 28-Aug-2018].

[57] TenneT TSO GmbH. "Herausforderung Bilanzkreisabrechnung: Was muss der Bilanzkoordinator leisten ? Rahmenprozesse zur Bilanzkreisabrechnung (Umsetzung MaBiS)." In: *BDEW-Infotag: Herausforderung BK-Abrechnung* (2010) . Available: https://www.tennet.eu/fileadmin/user_upload/The_Electricity_Market/Rules_and_procedures/ German/mabis/biko_tennet-vnb/Ausbilanzierung/bilanzierungsgebietsverwaltung-ausbilanzierung-regelzone.pdf. [Accessed: 03-Jan-2019].

[58] BDEW Bundesverband der Energie- und Wasserwirtschaft e. V. "EDI @ Energy Codeliste der Zeitreihentypen." (2017) . Available: https://www.bundesnetzagentur.de/DE/Service-Funktionen/Beschlusskammern/Beschlusskammer6/BK6_31_GPKE_und_GeLiGas/Mitteilung _Nr_58/Anlagen/Codeliste_Zeitreihentypen_1.1b.pdf?__blob=publicationFile&v=2. [Accessed: 08-Dec-2018].

[59] G. Fischer. "Bilanzkreismanagement Strom und Grundlagen MaBiS." In: *Seminar, proprietry at Campus-EW GmbH* (2018) .

[60] BDEW Bundesverband der Energie- und Wasserwirtschaft e. V. "MaBiS 3.0 Prozessbeschreibung auf Basis der BDEW-Position." (2018) . Available: https://www.bdew.de/media/documents/Stn_20180611_MaBiS_BDEW_Position.pdf. [Accessed: 06-Jan-2019].

[61] S. Haufe. "Bilanzkreiskoordination der Übertragungsnetzbetreiber." In: *7. Göttinger Energietagung, 50hertz* (2015) . Available: https://www.efzn.de/fileadmin/documents/Goettinger_Energietagung/Vorträge/2015/03_Haufe. pdf. [Accessed: 11-Dec-2018].

[62] J. Frunt. "Analysis of Balancing Requirements in Future Sustainable and Reliable Power Systems." Eindhoven University of Technology (2011).

[63] AG FPM and BNetzA. "Prozessbeschreibung Fahrplananmeldung in Deutschland." (2018) . Available: https://www.bundesnetzagentur.de/DE/Service-Funktionen/Beschlusskammern/1BK-Geschaeftszeichen-Datenbank/BK6-GZ/2018/2018_0001bis0999/BK6-18-061/BK6-18-061_konsult_prozessbeschreibung_fahrplananmeldung.pdf?__blob=publicationFile&v=2. [Accessed: 02-Jan-2019].

[64]   Bundesnetzagentur. "Be-schluss-kam-mer 6 Az.: BK6-18-032." (2018) . Available:
        https://www.bundesnetzagentur.de/DE/Service-Funktionen/Beschlusskammern/1BK-
        Geschaeftszeichen-Datenbank/BK6-GZ/2018/2018_0001bis0999/BK6-18-032/BK6-18-
        032_konsultation_basepage.html. [Accessed: 02-Jan-2019].

[65]   J. Gruber. "Germany - decentralized data management." In: *EnBW* (2015) . Available:
        https://www3.eurelectric.org/media/176895/gruber.pdf. [Accessed: 28-Nov-2018].

[66]   THEMA Consulting Group. "Data Exchange in Electric Power Systems: European State of
        Play and Perspectives." (2017) . Available:
        https://www.entsoe.eu/Documents/News/THEMA_Report_2017-03_web.pdf. [Accessed: 25-
        Nov-2018].

[67]   S. Geitner. "Bachelor thesis. Einsatzmöglichkeiten der Blockchain-Technologie in der
        Energiewirtschaft." Hochschule für Angewandte Wissenschaften Landshut (2018).

[68]   Consentec GmbH. "Description of load-frequency control concept and market for control
        reserves." (2014) . Available: http://www.consentec.de/wp-
        content/uploads/2014/08/Consentec_50Hertz_Regelleistungsmarkt_en__20140227.pdf.
        [Accessed: 22-Dec-2018].

[69]   ENTSO-E. "The Harmonised electricity role model VERSION: 2017-01 APPROVED." (2017)
        . Available: https://docstore.entsoe.eu/Documents/EDI/Library/HRM/2015-September-
        Harmonised-role-model-2015-01.pdf. [Accessed: 22-Sep-2018].

[70]   J. Kunkel. "Energiemengenbilanzierung – Chancen und Risiken der Energiehändler in der
        Versorgung." In: *7. Göttinger Tagung zu aktuellen Fragen zur Entwicklung der
        Energieversorgungsnetze* (2015) . Available:
        https://www.efzn.de/fileadmin/documents/Goettinger_Energietagung/Vorträge/2015/04_Kunke
        l.pdf. [Accessed: 04-Jan-2019].

[71]   T. Zimmermann, S. Klaiber, and P. Bretschneider. "A Market System Operator as a new role to
        balance the distribution grid and to coordinate market and grid operations." In: *2016 IEEE
        International Energy Conference, ENERGYCON 2016* (2016) .

[72]   www.netztransparenz.de. "Prozessbeschreibung Fahrplanabwicklung in Deutschland." (2018) .
        Available: https://www.netztransparenz.de/portals/1/Content/EU-Network-Codes/EB-
        Verordnung/20180227_Prozessbeschreibung_Fahrplanabwicklung_
        in_Deutschland_Version_4.0.pdf. [Accessed: 01-Dec-2018].

[73]   50hertz GmbH *et al.* "Method for determining the reBAP." (2018) . Available:
        https://www.regelleistung.net/ext/static/rebap. [Accessed: 19-Dec-2018].

[74]   Bundesnetzagentur. "Modell zur Berechnung des regelzonenübergreifenden einheitlichen
        Bilanzausgleichsenergiepreises (reBAP)." In: *BK6-12-024 Beschlusskammer 6* (2016) .
        Available:
        https://www.transnetbw.de/downloads/strommarkt/bilanzkreismanagement/Modellbeschreibun
        g_reBAP_ab_05_2016.pdf. [Accessed: 20-Dec-2018].

[75]   Bundesnetzagentur. "Festlegungsverfahren zur Weiterentwicklung der
        Ausschreibungsbedingungen und Veröffentlichungspflichten für Sekundärregelung und
        Minutenreserve, Eckpunktepapier mit Zusammenfassung der Stellungnahmen." (2015) .
        Available: https://www.bundesnetzagentur.de/DE/Service-

Funktionen/Beschlusskammern/1BK-Geschaeftszeichen-Datenbank/BK6-
GZ/2015/2015_0001bis0999/BK6-15-158/BK6-15-158_Verfahrenseroeffnung.html.
[Accessed: 20-Oct-2018].

[76]   Bundesnetzagentur. "SMARD Strommarktdaten." (2018) . Available:
        https://smard.de/home/downloadcenter/download_marktdaten/726#!?downloadAttributes=%7
        B%22selectedCategory%22:2,%22selectedSubCategory%22:6,%22selectedRegion%22:%22D
        E%22,%22from%22:1543791600000,%22to%22:1544741999999,%22selectedFileType%22:
        %22PDF%22%7D. [Accessed: 13-Dec-2018].

[77]   50hertz *et al.* "Regelleistung.net Internetplatform zur Vergabe von Regelleistung." (2018) .
        Available: https://www.regelleistung.net/ext/data/. [Accessed: 13-Dec-2018].

[78]   C. Sperling and J. Päffgen. "Preisbegrenzung auf dem Regelenergiemarkt: Hintergründe und
        Ursachen." (2018) . Available: https://www.next-kraftwerke.de/energie-blog/regelenergie-
        preisbegrenzung-9999-euro. [Accessed: 10-Dec-2018].

[79]   J. Aengenvoort, J. Päffgen, and C. Sperling. "Für eine Handvoll Dollar – High Noon am
        Regelenergiemarkt." (2018) . Available: https://www.next-kraftwerke.de/energie-blog/fur-
        eine-handvoll-dollar-high-noon-am-regelenergiemarkt. [Accessed: 20-Dec-2018].

[80]   M. Merz. "The EnerChain project." (2018) . Available: https://enerchain.ponton.de/.
        [Accessed: 19-Aug-2018].

[81]   W. Curth. "Bricklebrit Lastgangbepreisung." (2018) . Available:
        http://bricklebrit.com/rebap.html. [Accessed: 02-Sep-2018].

[82]   J. Hu *et al.* "Identifying barriers to large-scale integration of variable renewable electricity into
        the electricity market: A literature review of market design." In: *Renewable and Sustainable
        Energy Reviews* (2018) . Available: https://doi.org/10.1016/j.rser.2017.06.028. [Accessed: 29-
        Nov-2018].

[83]   C. Redl *et al.* "Refining Short-Term Electricity Markets to Enhance Flexibility." (2016) .

[84]   C. Weber and S. Just. "Strategic Behavior in the German Balancing Energy Mechanism:
        Incentives, Evidence, Costs and Solutions." In: *Journal of Regulatory Economics* (2012) .

[85]   V. Pflieger. "Bestimmtheitsmaß R^2 - Teil 2: Was ist das eigentlich, ein R^2?" In: *INWT statistics*
        (2014) . Available: https://www.inwt-statistics.de/blog-artikel-lesen/Bestimmtheitsmass_R2-
        Teil2.html. [Accessed: 16-Dec-2018].

[86]   H. Chao *et al.* "Design of Wholesale Electricity Markets." (1999) . Available:
        http://web.mit.edu/esd.126/www/StdMkt/ChaoWilson.pdf. [Accessed: 07-Dec-2018].

[87]   A. Bogensperger *et al.* "Die Blockchain-Technologie Chance zur Transformation Der
        Energiewirtschaft? Berichtsteil Anwedungsfälle." In: *Forschungsstelle für Energiewirtschaft
        e.V.* (2018) . Available:
        https://www.ffe.de/attachments/article/846/Blockchain_Teilbericht_UseCases.pdf. [Accessed:
        23-Dec-2018].

[88]   S. Just. "The German Market for System Reserve Capacity and Balancing Energy." In: *Energy
        Economics* (2015) .

[89]   R. Kuwahata. "Elia Grid International." (2017) . Available: https://www.ider-

project.jp/stage2/feature/00000175/20170420_Slides.pdf. [Accessed: 15-Nov-2018].

[90] 50hertz *et al.* "Inanspruchnahme Ausgleichsenergie des EEG-Bilanzkreises nach § 2 Nr. 6 EEAV." (2018) . Available: https://www.netztransparenz.de/EEG/Transparenzanforderungen/Inanspruchnahme-Ausgleichsenergie. [Accessed: 29-Sep-2018].

[91] Baringa Partners LLP. "Forward hedging under I-SEM." (2016) . Available: www.baringa.com. [Accessed: 11-Dec-2018].

[92] S. Franz. "Batteries , Bits and Business: Latest trends in the German energy transition." In: *Büro F* (2016) .

[93] Eurelectric. "Designing fair and equitable market rules for demand response aggregation." (2015) .

[94] H. Ziegler, T. Mennel, and C. Hülsen. "Demand Response Activation by Independent Aggregators As Proposed in the Draft Electricity Directive." In: *DNV-GL, Eurelectric* (2017) .

[95] SEDC. "Explicit Demand Response in Europe: Mapping the Markets 2017." (2017) . Available: https://www.smarten.eu/wp-content/uploads/2017/04/SEDC-Explicit-Demand-Response-in-Europe-Mapping-the-Markets-2017.pdf. [Accessed: 07-Jan-2018].

[96] BNetzA. "Verordnung über den Zugang zu Elektrizitätsversorgungsnetzen (Stromnetzzugangsverordnung - StromNZV)." (2006) .

[97] D. Geode and I. Einzelnen. "im Festlegungsverfahren zur Erbringung von Sekundärregelleistung und Minutenreserve durch Letztver- braucher gemäß § 26 a StromNZV – BK6-17-046." (2016) .

[98] A. Flamm. "The Demand Response business model and barriers in the German market." In: *EnerNOC - BSEC Seminar* (2014) . Available: https://www.diw.de/documents/dokumentenarchiv/17/diw_01.c.492564.de/bsec_flamm.pdf. [Accessed: 07-Nov-2018].

[99] Working Group of IndustRE. "Business models and market barriers." (2016) .

[100] R. Verhaegen and C. Dierckxsens. "Existing business models for renewable energy aggregators." (2016) . Available: http://bestres.eu/wp-content/uploads/2016/08/BestRES_Existing-business-models-for-RE-aggregators.pdf. [Accessed: 19-Oct-2018].

[101] NextKraftwerke. "Was ist ein Bilanzkreis?" (2018) . Available: https://www.next-kraftwerke.de/wissen/strommarkt/bilanzkreis. [Accessed: 09-Nov-2018].

[102] F. Hasse *et al.* "Blockchain – Chance für Energieverbraucher?" In: *PwC* (2016) .

[103] C. Schneider. "Legal Update Energy law: Blockchain in the energy sector." In: *greenmatch.ch* (2018) . Available: https://www.greenmatch.ch/en/blog/legal-update-blockchain. [Accessed: 24-Nov-2018].

[104] O. Lohmann. "The legal framework of blockchain in the energy industry." In: *Strommarkttreffen - digitalization of the energy economy* (2017) .

[105] BDEW Bundesverband der Energie- und Wasserwirtschaft e.V. "BDEW-Strompreisanalyse

2018." (2018) . Available: https://www.bdew.de/media/documents/1805018_BDEW-Strompreisanalyse-Mai-2018.pdf. [Accessed: 17-Dec-2018].

[106] E. Götz, A. Haber, and S. Hauke. "Blockchain als Lösungsansatz für die zukünftige Stromversorgung?" In: *Symposium Energieinnovation, Graz/Austria* (2018) . Available: https://www.researchgate.net/publication/323446733_BLOCKCHAIN_ALS_LOSUNGSANS ATZ_FUR_DIE_ZUKUNFTIGE_STROMVERSORGUNG. [Accessed: 14-Dec-2018].

[107] H. Wirth and K. Schneider. "Recent Facts about Photovoltaics in Germany." (2018) . Available: https://www.ise.fraunhofer.de/content/dam/ise/en/documents/publications/studies/recent-facts-about-photovoltaics-in-germany.pdf. [Accessed: 29-Dec-2018].

[108]   von J. Albersmann *et al.* "Die digitalisierte dezentrale Energieversorgung von morgen gestalten." In: *PwC* (2017) . Available: https://www.pwc.de/de/energiewirtschaft/studie-gestaltungsmoeglichkeiten-energieversorgung.pdf. [Accessed: 22-Dec-2018].

[109] E. Smole. "Blockchain in the energy sector." (2018) . Available: https://www.bdew.de/media/documents/Studie-Blockchain-englische-Fassung-Dez.2018.pdf. [Accessed: 21-Dec-2018].

[110] R. Berlet, A. von Jagwitz, and D. Jander. "Generalized Operational FLEXibility for Integrating Renewables in the Distribution Grid (GOFLEX); D10.2 Demand Side Management, Opportunities and Restrictions in the European Market." (2017).

[111] B. Scholtka and J. Martin. "Blockchain – Ein neues Modell für den Strommarkt der Zukunft?" In: *RdE – Recht der Energiewirtschaft* (2017) .

[112] P. Olivella-Rosell *et al.* "Optimization problem for meeting distribution system operator requests in local flexibility markets with distributed energy resources." In: *Applied Energy* (2018) .

[113] P. Pinson *et al.* "The Emergence of Consumer-centric Electricity Markets." (2017) . Available: http://pierrepinson.com/docs/pinsonetal17consumercentric.pdf. [Accessed: 01-Sep-2018].

[114] E. Sorin, L. Bobo, and P. Pinson. "Consensus-based approach to peer-to-peer electricity markets with product differentiation." (2018) . Available: http://arxiv.org/abs/1804.03521. [Accessed: 15-Dec-2018].

[115] Next Kraftwerke GmbH. "Balancing group management & portfolio management." (2018) . Available: https://www.next-kraftwerke.de/virtuelles-kraftwerk/stromhandel/bilanzkreismanagement. [Accessed: 24-Dec-2018].

[116] Neas Energy GmbH. "Bilanzkreismanagement & Portfoliomanagement." (2018) . Available: https://www.neasenergy.de/maerkte/der-deutsche-markt/. [Accessed: 24-Dec-2018].

[117] RheinEnergie AG. "Bilanzkreismanagement Strom. Bleiben Sie in Balance." (2018) . Available: https://www.rheinenergie.com/de/unternehmensportal/ueber_uns/rheinenergie_trading/produkt e_2/services_strom/detailseite_63.php. [Accessed: 24-Dec-2018].

[118] SOPTIM AG. "SOPTIM Energy für Vertrieb, Beschaffung & Handel." (2018) . Available: https://www.soptim.de/de/vertrieb-beschaffung-handel/soptim-energy/. [Accessed: 24-Dec-2018].

[119] Stadtwerke Schwäbisch Hall GmbH. "Dienstleistungen." (2018) . Available: https://www.stadtwerke-hall.de/geschaeftskunden/dienstleistung/. [Accessed: 03-Dec-2018].

[120] EnBW Energie Baden-Württemberg AG. "Was ist eigentlich Direktvermarktung?" (2018) . Available: https://www.enbw.com/geschaeftskunden/handelskunden/produkte/direktvermarktung/. [Accessed: 24-Dec-2018].

[121] D. Donnerer and S. Lacassagne. "Blockchains and energy transitions What challenges for cities?" In: *Energy Cities* (2018) . Available: http://www.energy-cities.eu/IMG/pdf/energy-cities-blockchain-study_2018_en.pdf. [Accessed: 07-Jan-2019].

[122] F. Moret and P. Pinson. "Energy Collectives: a Community and Fairness-based Approach to Future Electricity Markets." In: *IEEE Transactions on Power Systems* (2018) . Available: http://ieeexplore.ieee.org/document/8301552/. [Accessed: 29-Oct-2018].

[123] Lumenaza GmbH. "Lumenaza services." (2018) . Available: https://www.lumenaza.de/en/our-services/. [Accessed: 24-Dec-2018].

[124] P. Bronski *et al.* "The Decentralized Autonomous Area Agent (D3A) Market Model." In: *Energy Web Foundation* (2018) . Available: https://energyweb.org/wp-content/uploads/2018/04/EWF-D3A-ConceptBrief-FINAL201804.pdf. [Accessed: 07-Aug-2018].

[125] E. A. Jordan, K. Kusakana, and L. Bokopane. "Prospective architecture for local energy generation and distribution with Peer-to-Peer electricity sharing in a South African context." (2018) .

[126] E. Mengelkamp *et al.* "Trading on local energy markets: A comparison of market designs and bidding strategies." In: *International Conference on the European Energy Market, EEM* (2017)

.

[127] Wikipedia. "Stablecoin." (2018) . Available: https://en.wikipedia.org/wiki/Stablecoin. [Accessed: 24-Dec-2018].

[128] Ethereum Parity. "Parity Bridge - Wiki." In: *Parity Tech Documentation* (2018) . Available: https://wiki.parity.io/Bridge. [Accessed: 27-Dec-2018].

[129] G. James Wood. "Polkadot is a heterogeneous multi-chain technology." (2018) . Available: https://polkadot.network/#cover. [Accessed: 27-Dec-2018].

[130] "Cosmos." (2018) . Available: https://cosmos.network/. [Accessed: 22-Dec-2018].

[131] D. Kajpust. "Blockchain Scaling Solutions: Cosmos and Plasma." In: *Medium* (2018) . Available: https://medium.com/tendermint/blockchain-scaling-solutions-cosmos-and-plasma-b5ee09456f80. [Accessed: 22-Dec-2018].

[132] G. Greenspan, M. Rozantsev, and A. Gilbert. "MultiChain - open platform for building blockchains." In: *Coin Sciences Ltd* (2018) . Available: https://www.multichain.com/. [Accessed: 22-Dec-2018].

[133] readthedocs. "Solidity." In: *GitHub* (2018) . Available: https://solidity.readthedocs.io/en/develop/. [Accessed: 27-Dec-2018].

[134] F. Blom. "A Feasibility Study of Blockchain Technology As Local Energy Market

Infrastructure." Norwegian University of Science and Technology Department (2018).

[135] BDEW. "EDI@Energy-Dokumente." (2018) . Available: https://www.edi-energy.de/index.php?id=1&no_cache=1. [Accessed: 29-Nov-2018].

[136] M. Opray. "Could a blockchain-based electricity network change the energy market? | Guardian Sustainable Business | The Guardian." In: *The Guardian* (2017) . Available: https://www.theguardian.com/sustainable-business/2017/jul/13/could-a-blockchain-based-electricity-network-change-the-energy-market. [Accessed: 04-Dec-2018].

[137] BDEW Bundesverband der Energie- und Wasserwirtschaft. "Datenschutzerklärung des BDEW." . Available: https://www.bdew.de/datenschutzhinweise/. [Accessed: 27-Nov-2018].

[138] DSGVO. "Art. 17 DSGVO Recht auf Löschung ('Recht auf Vergessenwerden')." (2016) . Available: https://dsgvo-gesetz.de/art-17-dsgvo/. [Accessed: 05-Nov-2018].

[139] M. Schulz and J. A. Hennis-Plasscheart. "REGULATION (EU) 2016/679 OF THE EUROPEAN PARLIAMENT AND OF THE COUNCIL." (2016) . Available: https://eur-lex.europa.eu/legal-content/EN/TXT/?uri=CELEX:32016R0679. [Accessed: 25-Nov-2018].

[140] M. Buchhorn-Roth. "Blockchain and EDI for secure data exchange in supply chains." (2016) . Available: https://www.linkedin.com/pulse/blockchain-edi-secure-data-exchange-supply-chains-buchhorn-roth. [Accessed: 04-Dec-2018].

[141] Sky Republic. "How a Private Blockchain Can Enhance Your Existing EDI/API Infrastructure." . Available: https://skyrepublic.com/private-blockchain-can-enhance-existing-edi-api-infrastructure/. [Accessed: 04-Dec-2018].

[142] J. Lück. "Bilanzkreisvertrag, MaBiS und Ausgleichsenergie: Gestaltungsnotwendigkeiten und -herausforderungen der Regulierungsbehörde." In: *Göttinger Energietage* (2015) .

[143] Bundesnetzagentur. "Anlage 1 zum Beschluss BK6-07-002." (2012) . Available: https://www.bundesnetzagentur.de/DE/Service-Funktionen/Beschlusskammern/1_GZ/BK6-GZ/2007/2007_0001bis0999/2007_001bis099/BK6-07-002/BK6-07-002_Anlagen1und2zumBeschluss10062009_bf.pdf?__blob=publicationFile&v=4. [Accessed: 13-Oct-2018].

[144] Bilanzkreiskooperation and BNetzA. "Festlegungsverfahren der Bundesnetzagentur zur Änderung des Bilanzkreisvertrages Strom Sicht der Bilanzkreiskooperation." In: *Workshop der Bundesnetzagentur zu BK6-14-044* (2016) .

[145] S. Neumann, E. Demidova, and M. Kohlhoff. "Potenziale der Blockchain in der Energiewirtschaft." In: *ew Spezial* (2017) .

[146] B. Flieger *et al.* "Zukunftsfeld Mieterstrommodelle Potentiale von Mieterstrom in Deutschland mit einem Fokus auf Bürgerenergie." (2018) .

[147] H.-W. Schiffer and N. Kaim-Albers. "Energie für Deutschland - Fakten, Perspektiven und Positionen im globalen Kontext | 2017." In: *Weltenergierat – Deutschland e.V.* (2017) . Available: http://www.weltenergierat.de/wp-content/uploads/2014/02/Energie-für-Deutschland-2017_.pdf. [Accessed: 22-Oct-2018].

[148] T. Federico. "Blockchain und die Energiewirtschaft." In: *PV MAGAZINE* (2016) . Available: https://www.pv-magazine.de/2016/10/13/blockchain-und-die-energiewirtschaft/. [Accessed:

10-Oct-2018].

[149] BDEW Bundesverband der Energie- und Wasserwirtschaft e. V. "Blockchain in der Energiewirtschaft Potenziale für Energieversorger." (2017) . Available: https://www.bdew.de/media/documents/BDEW_Blockchain_Energiewirtschaft_10_2017.pdf. [Accessed: 27-Nov-2018].

[150] M. Hinterstocker *et al.* "Faster switching of energy suppliers – a blockchain-based approach." In: *Energy Informatics* (2018) . Available: https://energyinformatics.springeropen.com/articles/10.1186/s42162-018-0055-x. [Accessed: 12-Dec-2018].

[151] C. Dähne. "Bilanzkreis- und Fahrplanmanagement vor dem Hintergrund des Energiewirtschaftgesetzes mit der Stromnetzzugangsverordnung (StromNZV)." in *VBEW-Seminar* 2005 pp. 1–70.

YOUR KNOWLEDGE HAS VALUE

- We will publish your bachelor's and
  master's thesis, essays and papers

- Your own eBook and book -
  sold worldwide in all relevant shops

- Earn money with each sale

Upload your text at www.GRIN.com
and publish for free